SENTIENCE

The author (left), with (clockwise) Marvin Minsky, Stephen Jay Gould, Daniel Dennett, John Brockman, Eastover Farm, Connecticut, 1995
Photograph: Katinka Matson

Also by Nicholas Humphrey:

Consciousness Regained
The Inner Eye
In a Dark Time (ed. with Robert Lifton)
A History of the Mind
Leaps of Faith
The Mind Made Flesh
Seeing Red
Soul Dust

SENTIENCE

The Invention of Consciousness

NICHOLAS HUMPHREY

The MIT Press
Cambridge, Massachusetts
London, England

First MIT Press paperback edition, 2024

The MIT Press would like to thank the anonymous peer reviewers who provided comments on drafts of this book. The generous work of academic experts is essential for establishing the authority and quality of our publications. We acknowledge with gratitude the contributions of these otherwise uncredited readers.

This book was set in Albertina by Oxford University Press.
Printed and bound in the United States of America.

Library of Congress Cataloging-in-Publication Data is available.

ISBN: 978-0-262-04794-4 (hardcover)—978-0-262-54831-1 (paperback)

10 9 8 7 6 5 4

CONTENTS

Prologue vii

1. Sentience and consciousness 1
2. Foothills 13
3. The touch of light 16
4. Blythe spirits 21
5. What the frog's eye tells the monkey's brain 32
6. Blindsight 40
7. Sight unseen 50
8. Red sky at night 54
9. Nature's psychologists 67
10. On the track of sensations 77
11. Evolving sentience 101
12. The road taken 105
13. The phenomenal self 114
14. Theoretical misprisions 125
15. Coming to be: Sentience and body sense 130
16. Sentience all the way down? 134
17. Mapping the landscape 145
18. Getting warmer 148
19. Testing, testing 153
20. Qualiaphilia 158
21. The self in action 175
22. Taking stock 202

23. *Machina ex deo* 207
24. Ethical imperatives 214

Acknowledgements 219
References and notes 221
Index 239

PROLOGUE

Hello? (hello) (hello)

A place for big thoughts is a hot tub at night at the edge of the Mojave Desert. It's a couple of hours' drive from San Diego, where I've been at a meeting about human evolution. The lodge where I'm staying is surrounded by cactus-like trees. The Mormon pioneers called them Joshua trees because they seem to stretch their arms heavenwards. Lying on my back in the gurgling water, I gaze at the stars and sink into the vastness of space.

Is there anybody there?

If there are extra-terrestrial intelligent beings somewhere in our galaxy, they may be looking at the very same stars I am. Are they conscious of visual sensations like mine? Do they experience this phenomenal blackness, pricked with points of light?

I lie back, arms out sideways holding the rim of the tub. I feel the warmth of the water on my skin, smell the scent of the desert grasses. I'm flesh and blood. I'm soul. How can that be?

Hello, E.T., if you can hear me. Do you have this dual nature too? Is the light on inside your head? Do your sensations have the same eery immaterial quality that mine do?

I want to think so. I want this to be shared.

I hear a coyote bark, then another. Where else on Earth does sentience reside? Do dogs feel pain like mine? Does an earthworm enjoy smells? Are machines ever going to have conscious feelings? Do they already? How could we know?

Barking again. Have the coyotes caught a rabbit? Poor rabbit. One minute she's comfortably scratching her ear, the next a coyote has her by the neck.

What to say about the downside of sentience? The philosopher Schopenhauer wrote: 'If the reader wishes to see whether the pleasure in the world outweighs the pain, let him compare the respective feelings of two animals, one of whom is engaged in eating the other.'

There's a rock in the next valley, sculpted by the winds into the shape of a huge human skull. The hill in Jerusalem where Jesus was crucified was called Skull Rock—Golgotha in Aramaic, Calvary in Latin. Schopenhauer might have compared the feelings of two humans, one of whom is nailing another to a cross. During the crucifixion it's said that day turned to night. The stars came out.

But suppose conscious beings like us have not evolved anywhere else.

Suppose consciousness as it exists on Earth is a one-off accident of evolution. Astronaut Frank Borman, looking from the window of Apollo 8, remarked, 'the Earth is the only thing in the universe that has any colour'. This can't be strictly true. But it could be true that the Earth is the only place where sensations of colour exist. Or sensations of anything: sweetness, warmth, bitterness, pain. Which would be better: a universe without either joy or tears or a universe with both? Philosopher Thomas Metzinger agrees with Schopenhauer: the net utility is negative. He says that if an all-powerful and all-knowing 'superintelligence' could look across the world of pleasure and pain, and do the sums, it would conclude that it had a moral obligation to eliminate conscious life.

I think he's wrong. We don't live by bread alone. Pain and pleasure can't be *all* that matters. But there's no question *that* they matter. When we have reason to think someone is suffering, we have a duty of care. Some people think we have an equal duty towards any sentient being—human, nonhuman, even robot. It's not self-evident. But it could still be a rule we choose to live by. In that case, we have a heavy obligation to get it right about what in the world is conscious and what isn't.

The Norwegian government permitted the drawn-out, ugly, killing of more than 1,000 whales, including breeding females. But the Swiss government has made it illegal to boil lobsters alive and the British government soon may do the same.

Descartes believed that it's only humans who have feelings. Non-human animals are all unconscious machines. That's hard to believe. But maybe *some* animals are unconscious. A moth lands in the water of the hot tub. I scoop it out and toss it aside. Descartes could be right about moths. I hope he is right about moths.

Could Descartes be right about extra-terrestrials? What if the life-forms that exist out there are simply jumped-up insects? They might be ever so clever and still not have conscious feelings. I don't believe there's any necessary connection between intelligence and sentience.

Many people—famous philosophers among them—still don't get this. They think that if an octopus can solve a picture-puzzle that a four-year-old child would have trouble with, it probably has sensations something like ours.

Frans de Waal asks, 'Are we smart enough to know how smart animals are?' He writes a beautiful book, *Mama's Last Hug*. He's sure animals have feelings on the same level that we do. But what he comes up with as evidence is no more than a laundry list of clever tricks.

What about the Argument from Continuity? People say evolution has been a gradual process, without sharp discontinuities. There won't have been any point in history where we could draw a line: unconscious that side, conscious this. So, some kind of consciousness must go all the way down.

Panpsychists—'conscious everywhere' theorists—believe that consciousness is a basic property of physical matter. Even a teacup has a smidgeon of conscious feeling. Panpsychism seems to me a really bad idea. What would a smidgeon be like? *Whose* experience would it be?

It seems to me that consciousness must either be fully fledged or not there at all. That's certainly my experience of it. I move abruptly in and out consciousness when I wake from sleep or fall back into it again. Why not a similarly abrupt transition in the course of evolution: a critical point when all of a sudden our ancestors woke up, the lights came on? One small step for the brain; one giant leap for the mind?

There's a moving star, slowly arcing across the sky. No, not a star, a manmade satellite.

We human beings are on our way to becoming extra-terrestrial ourselves. Soon there will be sentient beings in space. But we, in our human bodies, won't be able to go beyond our solar system. If we want to explore the stars, we'll have to send intelligent robots in our place. Could these be sentient robots—machines that value their consciousness as we do? What extra ingredient would be required in their design?

Daniel Hillis has suggested that the World Wide Web has already become conscious simply as a consequence of its complexity. Only it hasn't deigned to tell us yet. Could the World Wide Web be hurting? Do we have a duty of care towards it?

It's often claimed that, as we build more and more complex robots, it will just happen: a threshold will be reached where sentience arrives as an emergent property—in the same way it happened in evolution. But I don't think it did just happen like that in evolution. I believe circuitry had to be built into the brains of our ancestors by natural selection for the special purpose of adding sentience.

There are many who disagree with me. They point out—quite rightly—that sentience can have been selected for only if it makes a positive difference to survival. But then, they ask, where's the evidence it makes any difference at all?

Any difference *at all*? In my own case, I want to say it makes all the difference in the world: the difference between *being me* and not being me! Yes, but wanting to say it doesn't make it true. I could be kidding myself. There are a good many people who maintain that I am kidding myself.

Hmm. This is going to need some turning round.

Back home in the UK, a new 'Animal Welfare Bill (Recognition of Sentience)' is under consideration by parliament. Clause 1 is the 'animal sentience' clause. The Secretary of State says that this will 'embed in UK statute the principle that animals are sentient beings, capable of feeling pain and pleasure'. I see that Sir Stephen Laws, former First Parliamentary Counsel, has commented at the Committee stage: 'It is fair to say that all the concepts in Clause 1 seem to be problematic in one way or another.'

He's right. It's a philosophical, scientific, ethical, and legal mess. As of now, we lack not only direct evidence but even agreed arguments as to how far consciousness extends. Arguably, the only sure case is our own, and every other is within reasonable doubt. Yet, all the time, we are obliged to act as if we know the answers.

Mary Oliver ends her lovely poem about whether stones, trees, and clouds have conscious feelings, 'Do Stones Feel?' by protesting that even if the world says it's not possible, she refuses to concur. 'Too terrible it would be, to be wrong.'

I understand her: the poet's refusal to bow to impossibility, the insistent tug of the 'What if?' What *if* stones feel? I may say I share the world's opinion about stones. I'm as certain as can be that they don't feel. But lobsters, octopuses?

Terrible to be wrong. Yes. But irresponsible not to be right—if we can only establish *what right is*. Let's suppose we can discover how conscious feeling is generated in the brain and how it shows up in animals' behaviour. Then perhaps we'll even be able to have a diagnostic test.

When Archimedes, in his bath, realized how he could test whether or not the king's crown was pure gold, he leapt out and ran naked through the streets of Syracuse. I lie back in the tub in the desert and wait for that Eureka moment.

Philosopher Jerry Fodor has said, 'We don't know, even to a first glimmer, how a brain (or anything else that is physical) could manage to be a locus of conscious experience. This is, surely, among the ultimate metaphysical mysteries; don't bet on anybody ever solving it.'

It seems it might be a long wait.

1

SENTIENCE AND CONSCIOUSNESS

I have used the terms 'sentience' and 'consciousness' liberally in the prologue without attending to their definition. Easy to do when you're lounging in a bath. Easy to do, actually, when you're sitting at your desk and writing an academic paper. I admit I sometimes find myself at cross purposes even with myself.

Given the seriousness of our topic, we can't afford to be slapdash about the language. So, from here on, I'll be more careful, at the expense of being more long-winded.

Let's start with the words 'sentient' and 'sentience'. The adjective 'sentient' came into use in the early seventeenth century to describe any creature—human or otherwise—that responds to sensory stimuli. But the meaning subsequently narrowed to put emphasis on the inner quality of the experience: what sensations *feel* like to the subject. And by the time sentience, the state of being sentient, was under discussion, notably in an 1839 book about cruelty to animals,[1] what was at issue was whether animals have experiences that feel to them the way *ours* do to us.

Thus sentience, in the first instance at least, gets its meaning by ostension. If we ask whether another creature is sentient, our understanding of the question rests on our being able to point to personal examples of *what it's like for us*.

Being sentient means having experiences *like this*: like the sensation of redness *we have* when we look at a poppy or the sensation of sweetness *we have* when we taste a sugar lump.

As scientists, however, we must step back from our first-person involvement. We must get a fix on what sensations are objectively about. I'll return to this over and over again as we go forward. But let's say for now that sensations are basically mental states—ideas—that track what's happening at our sense organs: light at our eyes, sound at our ears, scent at our nostrils, and so on.

They provide us, as subjects, with information about the quality of the sensory stimulus, its distribution and intensity, its bodily location, and—especially—how we evaluate it: the pain is in *my toe* and *horrible*; the red light is at *my eyes* and *stirs me up*.

But 'tracking' this information is only half the story. For, as we can each of us attest, sensations have a qualitative dimension that sets them apart from all other mental states and attitudes. There's something that our pains, smells, sights, and so on have in common that our thoughts, beliefs, wishes, and so on don't. For want of a better word, let's say it's something uniquely 'charming'.

We may not be sure exactly *what* this charming something is, but we can be sure *that* it is. Suppose you were to acquire a new kind of sense organ in addition to those you already have, for example, a sense organ that registers magnetic fields. Your magnetic sensations might be as different from visual sensations as visual are from tactile or auditory. But if they were to be in their own way similarly charming, you would recognize immediately that they were up there with the rest.

Now, when we ask about sentience in a non-human creature, the same applies. The creature's sensations need not correspond to ours in every way. The creature might indeed have sense organs that we don't have. But the quality of its sensations must be such that, if they *were* ours, we would recognize them as belonging to the same charmed class.

Philosophers do have another word than charming. They call this special quality 'phenomenal quality', and they call particular examples of it—such as phenomenal redness or phenomenal sweetness—'qualia'. Moreover, they say that, when we experience qualia, 'it's like something' to have the experience: it's like something to feel the pain of a bee sting, and it would—or could be—like something to have the sensation of magnetic north. Though none of these terms is ideal, I'll make use of them as others do. A creature is sentient if but only if it consciously experiences qualia—by virtue of which it becomes like something to be itself.

*

Let's turn to the term 'consciousness'. When we are aware of having experiences with phenomenal qualities, we can be said to be 'phenomenally conscious'. For many philosophers, phenomenal consciousness is the only kind of consciousness that really matters. David Chalmers, for example, says, 'I use "experience", "conscious experience", and "subjective experience" more or less interchangeably as synonyms for phenomenal consciousness.'[2]

However, here we do need to exercise some care. The word 'consciousness' has been in use a good deal longer than sentience or phenomenal consciousness, and along the way it has acquired a considerably wider reach, both in everyday speech and in the science of psychology. The oldest meaning, going back to classical times, has to do with self-knowledge: we say a person is conscious of a mental state when they *know* they have it. A modern meaning in cognitive science has to do with information processing: we say a state is conscious when its contents are available to a *global workspace* in the brain. Neither of these meanings restricts consciousness to states that have phenomenal quality.

Phenomenal consciousness is unquestionably a variety of consciousness. We know about our sensations. They influence our judgements and decisions. But, compared to other mental states of which we're conscious, sensations are clearly in a class of their own. To understand what's so special about them, we're going to have to tell the story of phenomenal consciousness separately from the story of consciousness in its entirety.

So, here at the start, let's take a closer look at the landscape, to see where phenomenal consciousness fits in.

*

We can begin with a simple definition, going back to the original meaning. Consciousness means having knowledge of what's in your mind. Your conscious mental states comprise just those states to which at any one time *you have introspective access* and of which *you are the subject.*

These can include all sorts of mental states: memories, emotions, wishes, thoughts, feelings, and so on. When you introspect, you observe these various states, as it were, with an inner eye. Thus, it comes naturally—and people everywhere do this—for you to think of consciousness as some kind of window on the mind, a private view of the stage where your mental life is being played out.

A view from whose standpoint? Well, from the standpoint of whom else but 'You', your conscious *self*. Wherever your self focuses its gaze, it takes over as the singular subject of these states. And this speaks to one of the most striking features of consciousness: its *unity*. Across different states and across time, the conscious subject remains one and the same. There's only one 'You' at the window, only one self. When you find yourself feeling pain, or wanting breakfast, or remembering your mother's face, it's the same you in each case.

We might think it obvious it has to be so. But actually this unity is by no means a logical necessity. It's quite conceivable—and indeed psychologically plausible—that your brain could house several independent selves, each representing a different segment of the mind. In fact, this fragmented state may have been the way you started out at birth. Thankfully, however, it was never going to stay that way. As your life got going and your body—your one body—began interacting with the outside world, these separate selves were destined to come into register, orchestrated, as it were, by the single line of music that made up your life.

The unity of the self underwrites the most obvious cognitive function of consciousness, which is to create what Marvin Minsky has called the 'society of mind'. Just as—in fact just because—there is only one 'You' at the window looking in, there comes to be only one mind on the other side. Anything that is *in consciousness* becomes shareable with whatever else is. Information from different agencies is being brought to the same table, and it's here that your sub-selves can meet up, shake hands, and engage in fertile cross-talk. This means you now have a mind-wide forum for planning and decision making—a conscious workspace wherein you can recognize patterns, marry past and future, assign priorities, and so on. A computer engineer might recognize this as an 'expert system', designed to anticipate your environment and make intelligent choices. You, of course, recognize it as 'You'.

Then, alongside this, a different kind of opportunity emerges. Once you can observe the parts of the mind interacting on a single stage, you are in a position to *make sense* of the interaction and track its history. Observing, for example, how 'beliefs' and 'desires' generate 'wishes' that lead to 'actions', you find your mind revealed as having a clear psychological structure. Thus, you begin to gain insight into *why* you think and act the way you do. This means you can explain yourself to yourself and explain yourself to other people too. But, equally importantly, it means you have a model for explaining other people to yourself. When you meet another person, you can assume that their mind works much as yours does. So you can work out what they are likely to be thinking and how they will behave. Consciousness has laid the ground for what psychologists call 'Theory of Mind'.

To summarize: consciousness transforms how your mind works on two levels: (a) it creates a cognitive workspace, which makes you more intelligent; (b) it underwrites a coherent self-narrative, which helps you make sense of your own and others' behaviour.

*

You'll note that, up to this point, we've not assigned any special role to phenomenal experience. Now let's ask: where do sensations with phenomenal qualities fit in?

Sensations have much in common with other conscious states. You have access to them by introspection, and your unified self is their subject. They are readily available within the workspace, and the information they carry about sensory stimuli plays an important part in your mind's expert calculations. Moreover, sensations play a key role in your self-narrative.

But here's the puzzle. Sensations could play these roles even if they did not have the extra phenomenal quality. Nothing says phenomenal quality is *necessary*, that the information about sensory stimuli could not be utilized without this quality, nor that the only kind of useful self-narrative is one that centres on a phenomenally conscious self.

So, the immediate answer as to where phenomenal consciousness comes in would seem to be that it does not have to come in at all. The advantages of having a conscious self could quite well accrue to a creature whose sensations have none of the quality we humans take for granted. A creature lacking this dimension of experience could be capable of introspection, could know its own mind, have a self-narrative, be highly intelligent, goal-directed, motivated, percipient, and so on.

Such a creature, by our definitions, would undoubtedly possess a form of consciousness. Yet, because it would not be experiencing sensory qualia, it would not be sentient.

Of course, this possibility is hard for us humans to take on board. In so far as we can imagine it, consciousness in the absence of phenomenal experience might seem to be such an anaemic kind of consciousness that (along with David Chalmers) we'd question whether to call it real 'consciousness' at all. However, if as scientists we're interested in what consciousness can achieve at the level of mental organization, it's clear we ought to. Suppose we were to come across a creature that demonstrably possesses the cognitive faculties listed in the previous paragraph without any added phenomenality—in short, an insentient creature that walks like a conscious creature, swims like a conscious creature, and quacks like a conscious creature: then how could we not say that it really *is* a conscious creature?

But I agree this may not sound right. It's surely not *our kind* of conscious creature. So, to reduce ambiguity, and as a reminder of what may be missing even in a conscious duck, I suggest that when we're talking about consciousness as the umbrella term for introspectively accessible mental states, and as a mediator of cognitive operations of the kind just listed, we call this 'cognitive consciousness'. But when we're talking specifically about access to sensations that have phenomenal sensory qualities, we call this 'phenomenal consciousness'.

*

While recommending these terms, I should make a philosophical aside. The philosopher Ned Block, in an influential paper in

1995, argued that we should draw a distinction between what he called 'access consciousness' and 'phenomenal consciousness'.[3] This may sound exactly like the distinction I've suggested. But it's not. And I want to dissociate myself from it.

By phenomenal consciousness, Block did indeed mean the experience of sensations with phenomenal properties, or qualia; but he contended that qualia are walled off from the rest of the mind and play no part in guiding thought, speech, or action. Yet, to me, this never made much sense. Block was arguing in effect against the unity of consciousness: that the *You* who is the subject of qualia is a different You from the You who is the subject of all your other mental states. Not only is this counter-intuitive, but it also misses the point about what really make qualia distinctive: it's not their mode of access but their content—*what they are like.*

Consider this as an analogy. You have a library of books that you can take from the shelf. All the books have written texts, but a subclass also have picture illustrations. At any one time, you'll have some of the books open on your desk and you'll be browsing them, cross-comparing them, and so on. The open books are all equally accessible. However, the *qualitative content* of the picture books sets them apart—and makes you value them differently from the plain word books.

Now, to relate this analogy to our two kinds of consciousness, suppose sensations with phenomenal properties correspond to the picture books. Then you'll be cognitively conscious when you are browsing any kind of open book, but you'll be phenomenally conscious only if you're browsing a picture book.

As it happens, we humans rarely find ourselves in the situation of being cognitively conscious without being phenomenally conscious. While we are awake, we are always having some sort of sensations with phenomenal properties—so we always have books with pictures open on our desks. It's true that there are some remarkable exceptions where the phenomenality is missing: the state of sleepwalking would seem to be such a case,[4] and there are related conditions that result from brain damage—notably 'blindsight'—that we'll discuss in a later chapter. But, here's the tantalizing possibility: even if this is rarely the case for humans, perhaps there could be other animals for which it's always the case.

Scientists don't yet know when, in the course of evolution, sensations acquired phenomenal properties: when, as it were, the picture book was invented. That's what we want to discover. But I'll stick my neck out right away. I think it's quite possible that phenomenal consciousness arrived relatively late and long after cognitive consciousness was already in place. If that's so, for much of history, our ancestors could have been cognitively conscious but not phenomenally conscious—conscious but insentient. And presumably the same could still be true of many animals today.

The question is: how could we possibly tell from an animal's behaviour which side of the divide it is?

*

The octopus cracks a combination lock to escape from a box. The crow plans ahead to make sure it has something for breakfast. The chimpanzee outperforms humans on a memory task. These are almost certainly evidence of cognitive consciousness at work. But intellectual feats like these have no direct bearing on phenomenal consciousness. Engineers will soon have built something like cognitive consciousness into intelligent machines (if they haven't already), and these machines will be cleverer than any of us…, So what?

Back in 1820, the philosopher Jeremy Bentham put it this way: 'The question is not, Can they reason? nor, Can they talk? but,

Can they suffer?'[5] To bring this up to date with our more general concern: 'The question is not, Do they have a global workspace or a self-narrative, but are they sentient?' If signs of cognitive consciousness aren't going to be sufficient to provide the answer, what would be?

The astronomer Carl Sagan remarked that 'extraordinary claims require extraordinary evidence'. The claim that any creature—human or otherwise—is sentient is about as extraordinary a claim as we can make. It's an extraordinary claim even to make about ourselves. *'The one thing I know is I am sentient.'* In our own case, it seems we do indeed have extraordinary evidence: we are directly acquainted with the phenomenal properties of sensations by introspection. The truth of our claim is transparent to us in a way it isn't to anyone else. The scientist, however, who hopes to study sentience from the outside, is bound to ask: if they can't hope to feel someone else's sensations for themselves, is there some other *public* evidence from which they might be able to *deduce* what the experience is like?

I'll declare that I think there has to be. This is an article of faith. But it's supported by one very powerful argument: the evolutionary one. We know that at some point in history sentience came into the world as a remarkable internal feature of the minds of the animals from whom humans and other sentient species are descended. We also have very strong grounds to believe that there is only one way by which new species-wide features can emerge and become stabilized in the course of biological evolution and that is by *natural selection*—the process, discovered by Darwin, by which heritable traits spread through a population if they help their owners win out in the struggle to survive and reproduce. And it stands to reason that, for a feature to be selected in this way, it must be having effects of some sort in the public realm.

At the beginning of this chapter, I said that we each define the *idea* of sentience by pointing to our own private experience. What I'm now saying is that this cannot be how natural selection recognizes the *fact* of sentience. You and your new trait don't get to survive better if all you can do is point to it in private. You must have something to show for it on the outside that natural selection can latch on to—something that affects biological survival.

This doesn't mean that, in order for it to have been selected, your experience must be as accessible to others as it is to you. You don't have to wear your experience on your sleeve for all to see. But your private experience must indeed have closely coupled public consequences that can be 'seen' by natural selection. And, if natural selection can see these consequences, presumably they must be seeable by other kinds of outside observers—scientists, philosophers, poets?—if only they know what to look for.

It's this consideration that will, I hope, give teeth to our enquiry into who else besides humans are sentient. It should give us the courage—perhaps against our own intuitions—to try to follow in nature's footsteps and establish *the face value of sentience*. We can try to discover—or perhaps work out from first principles—how phenomenal consciousness changes the subject's mental outlook, on some plane other than the purely intellectual one and in a way that impacts survival.

But, at the same time, we had better come up with a plausible story to explain how phenomenal consciousness could possibly be generated by a physically constituted brain and body. For, without a physicalist explanation, we'll be dogged by the philosophical nay-sayers—and there are lots of them out there—who would prefer to believe that sentience has its origin outside of physics and didn't actually evolve as a biological phenomenon.

*

Let me pause here. These are difficult questions. I have been asking them for fifty years. And I may say I've been ploughing a rather lonely furrow. Few of my colleagues have seen the issues in quite the way I do or have been looking for answers in the places I have.

We each approach questions like these in our own way, conditioned by our 'mental set'—our pre-existing framework of concepts and beliefs. And however objective we try to be, this is going to depend on the ideas and examples we've been exposed to along the road. In a classic study of mental set, people were shown the ambiguous rat/man drawing at the centre of Figure 1.1, after being primed—softened up, we might say—by exposure to drawings of either humans or animals. Come at this central figure from the left and you'll see a rat; come at it from the right and you'll see the man.

Figure 1.1

Similar biases operate at the level of theories. Come to the problem of the origin of species after reading *Genesis* and you'll see the working of 'intelligent design', but come to it after a voyage on the Beagle and you'll see natural selection. And now, come to the problem of sentience from neuroscience—or Buddhism...or evolutionary psychology...or chicken farming...or reading Beatrix Potter—and your answers are going to be different every time.

Come to the problem from where I've come from? Well, I want to tell you a bit more about that.

2

FOOTHILLS

I spent a whole day meditating—I should have done better to learn; I stood on tiptoe for a good view—better had I climbed a hill.

Chinese saying

The philosopher Daniel Dennett, in a review of my 2006 book, *Seeing Red*, remarked that, early in my career, I had some 'profoundly unusual encounters with the ill-understood boundaries of sight and experience—if only all who work on consciousness could have such intellectual adventures!' But, he went on, 'It is well nigh impossible to answer the tantalizing question: did they teach Humphrey important things the rest of us still find hard to imagine or take seriously, or did they seduce him down a theoretical cul de sac?'[6]

Dan is the cleverest person I know and one of the most generous. But it's tough being a philosopher: obliged to stand on tiptoe, while others, no more keen than you, have been climbing hills and encountering those unusually good views. In another world, Dan himself would have been a brilliant scientist. In this world, he has seen better than anyone how to make use of other scientists' findings in exploring the philosophy of mind.

When he invited me to come to work with him at Tufts University in 1988, the deal was that I would help keep him abreast of new research while he would help me think as a philosopher by closing off bad lines of argument. However, our situations were not symmetric. In the course of my own research, first in neuroscience, then in animal behaviour, I had seen things that changed my perspective—more than once—on what the problem of consciousness amounts to. Dan, on the other hand, still saw things very much the way he always had done since he set out his stall in his PhD thesis in 1965. Dan was rather proud of having stuck to his guns. I was proud of being a turncoat.

We made many long road trips between Boston and his farm in Maine. I would regale him with stories of blind monkeys who could see, gorillas who could read minds, humans who communicate with the dead. Dan would counter with yeses and buts. I'm glad to say that, years later, we have come to see eye to eye on many of the issues. Yet, he still finds some of the things I rely on as the base for discussing consciousness 'hard to imagine or take seriously'. And in that review Dan claimed that it's not just him that doesn't get it, it's 'the rest of us'.

If I'm to carry you with me in this book, I need to change this. To that end, the next several chapters are about first encounters. I'm going to introduce you to some of the people and animals and experiments I had the good fortune to meet with in the morning of my career—and I'll explain how they planted the ideas I'm still pursuing fifty years later.

*

I know this narrative structure is something of a risk. You may wonder where it's leading. You may think there's too much of me in it. I can only hope the value of telling it as a personal history will become clear as we go on. I've another thing to say up front about the book's arrangement. By the end, I hope we'll have answers to the question of who or what in the world—besides our human selves—could possibly be sentient. To get there we'll need a theory of how and why this remarkable form of experience could have come into existence as a biological phenomenon that was favoured by natural selection. Once we get to the theory, I'm going to be introducing some difficult ideas. None of it will be technical or require prior knowledge on your part. But, as you may already have found in Chapter 1, there will be passages where what I'm saying is counter-intuitive—where you'll have to think twice before agreeing with it (and, of course, on second thoughts, may discover that you don't!).

There's going to be give and take between stories, facts, and theory. I'll let facts drive the quest for explanations, and explanations drive the search for facts. Think of it as a performance on the concertina. As the bellows are pulled out, fresh air flows inwards. As they are pushed in, this same air flows outwards. The reeds pick it up in each direction. So we'll have questions followed by answers, answers by questions—and, hopefully, a line of argument that's all of a piece.

3

THE TOUCH OF LIGHT

I arrived as a student at Cambridge University in 1961 and was soon thrown in at the deep end. My tutor in physiology, Giles Brindley, sent me a note asking me to meet him in his lab. I went at the appointed time and knocked on the door. A distant voice said 'Enter'. I opened the door onto a room in darkness. 'Over here', said the voice. In the far corner, I could make out the dim glow of a cathode ray tube. 'You can turn on the room lights.' I now saw him. He was naked except for a pair of shorts and was standing in a bath of salt water. On his head was a helmet from which a metal rod projected against the side of his right eye. He held an electrical switch button in his hand.

'Humphrey, is it? You're early.' But I wasn't early. He had obviously planned that I should walk in on him in just this state. 'Never mind. You can see my experimental set-up. I'm following up on Newton's study of phosphenes. When I press the button the current flows from the rod through the back of the retina.' He pressed the button. 'There, I still see the annulus with the room lights on. But now the effect is the reverse of what it was in the dark, not red but blue.' 'What do you mean, you *see* it?', I wanted to ask. How can you see electric current? But I let it go. As he put his clothes on, he opened up. 'I expect you'll want to be an experimental subject yourself. I tell my students, it's facts that count, not theories. Of course the most telling facts are those you observe first-hand.' Yes, count me in.

Brindley, as I would find out later, had something of a reputation for self-experiment. In a classic study of the phenomenon of 'double pain', he showed that when someone receives a nasty electric shock to his foot, he feels two separate pain sensations, one almost immediately—carried to the brain by fast nerve fibres—and another two or three seconds later—carried by much slower fibres. He himself was the main experimental animal.

He had a taste for theatre too. In 1983, he went on stage at an international conference about erectile dysfunction sporting an erection clearly visible under his trousers. He explained to the audience that he had injected himself with papaverine in his hotel room before coming to give the lecture. As a gobsmacked witness described it later:

> He then summarily dropped his trousers and shorts... He paused, and seemed to ponder his next move. The sense of drama in the room was palpable. He then said, with gravity, 'I'd like to give some of the audience the opportunity to confirm the degree of tumescence.' With his pants at his knees, he waddled down the stairs, approaching (to their horror) the urologists and their partners in the front row.[7]

But, back to phosphenes. Phosphenes are the visual phenomena that can be produced by stimulating the retina of the eye, not by light but by mechanical pressure or, as here, directly by electric current. They have been commented on since ancient times. Newton undertook his investigation in the 1670s at the age of twenty-four. In his rooms in Trinity College, he did a set of experiments that beats even Brindley's for self-sacrificing devotion to science. These are his notes:

> I tooke a bodkin [a large ivory pin] & put it betwixt my eye & the bone as neare to the backside of my eye as I could: & pressing my eye with the end of it there appeared severall white darke & coloured circles... If the experiment were done in a light roome so that though my eyes were shut some light would get through their lidds There appeared a greate broade blewish darke circle outmost... But on the contrary if I tryed the Experiment in very darke roome the circle apeared of a Reddish light.[8]

'Vision', he concluded, 'is made in the retina because colours are made by pressing the bakside of the eye.'

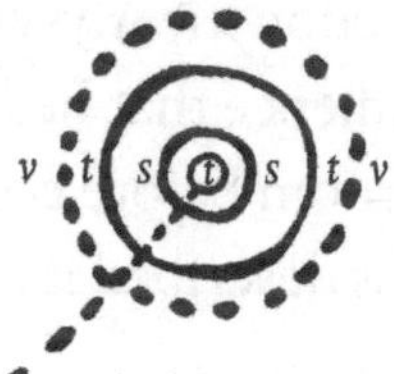

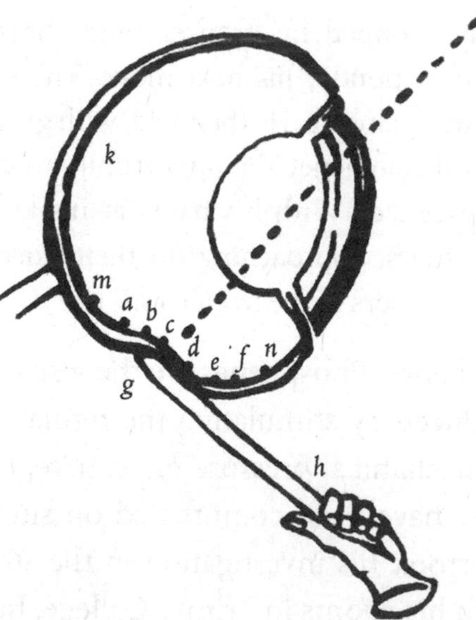

Figure 3.1 Newton's sketch of his experiment with phosphenes Laboratory notebook 1665, Cambridge University Library.

A week later, I was back in Brindley's lab with the helmet on my head and the rod against my eye. Yes, I saw it. The sensation was certainly visual. I knew at once why I should be interested in phosphenes. They reveal the fleshy side of vision.

Of our five senses, vision is often considered the loftiest, followed by hearing, smell, taste, and touch. Plato ranked the senses according to how far their reach goes beyond our bodies. Vision can tell us about the stars, touch can tell us only about what's in contact with our skin. Vision is therefore the least sullied by our animal nature.

But phosphenes bring vision back to earth. Although they are indeed visual in quality, they remind us that the retina is part of the skin. We feel the sensation created by stimulating the back of the eye in much the same way we would feel a tactile sensation created by touching the back of the hand. In fact, the similarity exists at the level of anatomy too. The photoreceptors in the retina, the rods, and cones, are modified sensory hairs repurposed by evolution to respond to the touch of light.

Phosphenes are sensations located at the eye. We experience them close in as things happening to our bodily selves. They may be as near as we'll ever get to experiencing the skin vision of our primordial ancestors. Vision *can* tell us about the stars. But that's visual *perception*, not *sensation*. Visual sensation is just as dirty and corporeal as sensations of smell, taste, or touch.

Rootling in the library, I discovered the writings of the eighteenth-century Scottish philosopher Thomas Reid. Reid insisted that we understand sensation not as a subordinate phenomenon, merely a stepping stone to perception, but as a significant phenomenon in its own right.

> The external senses have a double province—to make us feel, and to make us perceive. They furnish us with a variety of

> sensations, some pleasant, others painful, and others indifferent; at the same time they give us a conception and an invincible belief of the existence of external objects.[9]

This rang home. I'd noticed for myself the two sides to vision: the fact that it can tell you, on the one hand, about the light arriving at your eye, the retinal image, and, on the other hand, about the existence of things out there in the world. In fact, I'd enjoyed pushing my eyes around to put these two representations—subjective sensation and objective perception—at cross purposes. I remember a chemistry lesson at school taught by Mr Crumpler: 'You're in my power, Crumpler. If I press on my right eye I can move my image of you to the bunsen burner that I'm perceiving with my left.'

'Things so different in their nature [as sensation and perception] ought to be distinguished', Reid wrote. Yes.

4

BLYTHE SPIRITS

I joined the student branch of the Society for Psychical Research. I soon became friends with the grand old man of parapsychology, the philosopher Charlie Broad. We would meet for tea in his rooms in Trinity, and we would talk about ghosts.

Broad lived in the same rooms in college that had been Newton's, and his favourite spot for talking and for tea was an armchair placed beside the window where once Newton caught a beam of sunlight in his prism and spread it like a rainbow on the floor. Seated in this chair one afternoon, Broad told me of his fears that the spirit world in the mid-twentieth century was losing all its colour. Not that spirits as such had finally gone to rest (new reports of hauntings, poltergeists, and so on reached him every day), but it seemed that these modern spirits no longer cut the dash they used to do. Their activities were becoming—dare he say it—increasingly vulgar. Only the previous day, he had heard of a poltergeist which was shifting caravans around a holiday camp near Great Yarmouth. If this trend continued, Banquo would soon be advertising tartans on the television and Hamlet's father taking coach parties round Elsinore.

The old philosopher's face fell as he contemplated the decline in standards. I could understand only too well why he had concluded his celebrated *Lectures on Psychical Research* by saying 'For my own part I should be more annoyed than surprised if I should find myself in some sense persisting immediately after the death of my present body.'[10]

Nevertheless, Broad, like Brindley, believed in first-hand observation. He swallowed his dismay and, the following week, he and I took the train to Great Yarmouth on the Norfolk coast. He was aged nearly eighty, I was twenty. We spent a cold and uncomfortable night in one of the caravans. To no avail. The caravans remained resolutely in place.

He shyly explained to me that, while this was disappointing, it was not unexpected. Unfortunately, it seemed he personally had a dampening effect on paranormal phenomena, so that in his presence they hardly ever happened. This was a problem that had dogged him all along. He was a 'psi-inhibitor'—someone who, because he adopted too rational an attitude, scared the spirits off.

Still, he advised me not to lose hope. And he had a plan for me. He was in touch with an English gentleman, Hugh Sartorius Whitaker, living on the island of Elba, who for many years had been receiving messages from the dead.

*

Whitaker was from an aristocratic family with roots in the Italian wine business in Sicily. His great grandfather was said to have invented Marsala. As a young man he had lived the life of a monied dandy, with fascist tendencies (one of his proudest moments had been when Mussolini had accepted a gift of his speedboat). But in the late 1930s, fuelled by an encounter with a parish priest who was passionate about dowsing, he turned to spiritualism. In a chapel on his estate, he began to make contact with heavenly beings through a spirit guide who revealed himself to be Agresara, a Tibetan monk.

Agresara urged him to record his teachings, which he would receive in the form of automatic writing. Whitaker welcomed this as his life's work. He became Agresara's scribe and arranged for the publication of three volumes of the teachings.[11] As the publisher announced, these ranged over 'a wide field, from the origin and nature of God, to life in Spirit, faith, prayer, Christianity, the after-life, reincarnation, the power of thought. spiritual evolution and other related subjects presented to meet the needs of humanity in this critical age'.

Through Broad's good offices, Whitaker now issued an invitation to the Cambridge Society to send a delegation to investigate him. Thus, it came about that in the summer of 1963 I and two friends drove from Cambridge in an old Triumph Herald to spend a week with Whitaker at Villa Il Tasso on Elba. We camped in the Alps and on a beach of Lake Trasimeno. We took the ferry to Elba and made it via mountainous roads to the Villa, surrounded by woods, perched above the sea on the north coast.

Whitaker proved to be the perfect host. A white-haired gentleman in his seventies with rather guarded eyes and a distinctly fussy demeanour. He welcomed us as pilgrims to the shrine and asked us to open our minds to whatever we would see.

Unfortunately, on arriving, I had more pressing concerns. I had picked up a bug on the way and was having tummy pains. My health problem helped break the ice. Whitaker immediately set about a diagnosis. He wrote down several alternatives on slips of paper and then held over them a platinum ball on a chain. It swung towards dysentery. He wrote down the names of several medications. It swung towards codeine. This was the way he always consulted the powers that be. I was impressed and soon felt better.

The serious business began next morning. We were invited to attend on him in his study at five to nine. He was sitting at his desk, pen in hand, as he did every day. At nine o'clock precisely he closed his eyes, uttered a few words of invocation, entered a trance, and started writing. He wrote continuously for the next hour. Then he stopped abruptly and pulled himself together. 'All right boys, let's move on. We're going on a picnic.'

The picnic was something. We travelled in a Rolls Royce to a favourite spot in the mountains. Another car followed with food and wine. After a lovely walk, we returned to find lunch laid out on a cloth under a shady chestnut tree. It seemed an opportune time to make a start on the research which was our excuse for being there. So I ventured to ask our host what he had written about that morning. 'My secretary will type it up. Then it will become apparent. Now, don't ask questions, you'll spoil things.'

I soon realized that, for Whitaker, the point of having us visit was not so much that we should investigate his powers as be a witness to them. The deeper meaning of the writings would probably have been beyond our fathom anyway. Here is a sample, taken from Volume I of the 'Teachings'.

> Certain mysterious happenings cannot so far be explained by scientific knowledge you can guess what this mysterious agent is that is operated by the Superlatively Supreme Intelligence. But when you are asked where Spirit is located, and how it is formed, it is difficult for any of you to answer those questions: unless by practice you have succeeded in actually identifying yourself with this Spirit, by having temporarily surrendered that human conception of being an independent entity, so as to become part of that far Superior One, which actually comprises every human quality, and likewise, every other quality in animal, bird, fish, and everything that grows.

When I did get a peek at the typescript of one of the sessions we attended, I saw that Agresara also took an interest in contemporary politics: he had scathing things to say about Britain's Prime Minister, Harold Wilson, and the evils of socialism.

We weren't treated to any great revelations in the ensuing days. This was certainly a more agreeable place to be ghost hunting than a windswept caravan site. But the spirits remained in hiding. Nonetheless, I came away wiser than I arrived. I had been confronted with a prime example of just how weird the human mind is. Or, rather, how weird the human mind's conception of the mind is.

*

What kind of mind believes this kind of stuff? *Our kind of mind.* I was coming to see that there's something dreamily mad about human consciousness. It gives us a wonderfully inflated notion of our own metaphysical importance. Whitaker genuinely believed he was channelling a dead Tibetan monk. Professor Broad believed that the spirits of angry children could move caravans. Humans everywhere believe in telepathy, clairvoyance, precognition. Above all, they believe that the mind is more than matter.

Broad's classic work of philosophy was a tome called *The Mind and Its Place in Nature.*[12] In it, he argued that the mind is partly material and partly psychic. He was impressed by the evidence gathered by the Society for Psychical Research for the persistence of the human mind after death. What he found particularly convincing were spiritualist seances where a dead person, contacted by a medium, proved able to disclose information presumably unknown to anyone else. Yet, careful thinker that he was, he had a major reservation. He couldn't but remark that most messages from the dead were decidedly off-key, as if after death the mind becomes degraded morally and intellectually. Not to put too fine a point on it, the messages were mostly twaddle. He concluded, therefore, that the part of the mind that survives death—the 'psychic factor'—is less than the full ego of the living person, although it has been shaped by the person's life experiences.

This all struck me as crazy. But it also set me thinking. Could it be that consciousness has actually been designed by evolution to give us this larger-than-life conception of ourselves? In doing so, does it leave us vulnerable to extravagant myths about our own metaphysical importance, ready to clutch at whatever straws are on offer?

If so, is it just humans? Could non-human animals—if they are sentient in the way humans are—be similarly awestruck by their own existence? These may have been naive questions. But they would stay with me.

*

I held on to my undergraduate interest in psychical research. I liked the frisson of unrespectability that attached to it. I was struck by what Broad had written in the Preface to his *Mind* book:

> I shall no doubt be blamed by certain scientists, and, I am afraid by some philosophers, for having taken serious account of the alleged facts which are investigated by Psychical Researchers. I am wholly impenitent about this. The scientists in question seem to me to confuse the Author of Nature with the Editor of *Nature*; or at any rate to suppose that there can be no productions of the former which would not be accepted for publication by the latter. And I see no reason to believe this.[13]

I had every intention of having a career in science, but I certainly didn't want to be one of *those* scientists: someone who assumed that the laws of nature, as known today, would never have to be revised. Nonetheless, I was generally sceptical about what Broad called 'the alleged facts'. Not that they weren't fascinating. But, from what I'd read about them, and the little I'd seen, I reckoned that the fact of the *allegation* was often of more interest than the fact of the facts.

I was a psychology student, not a philosopher. I wanted to understand why people are so ready to believe in things that probably aren't true: in particular, why they are so keen to believe in the supernatural origins of natural events that they actually find psychical explanations more congenial than physical ones. My larger agendum was to understand how consciousness itself may be deceiving us. I wondered whether the study of other examples of human gullibility might throw light on the evolutionary trajectory that consciousness has taken.

*

As I started on a PhD, I had more conventional scientific questions to pursue. But a few years later, in a break between jobs, I was able to follow up with some hands-on investigations of alleged paranormal events. In 1985, I was offered the opportunity to make a television documentary about belief in the paranormal.[14] The events we looked at for this programme were all, in one way or another, *illusions of supernatural intervention.*

I realized that there were two strikingly different pathways by which such illusions could be generated: by accident or by design. For the documentary programme, we looked into two contrasting examples, both of which took place in Ireland. I'll describe them here for the bearing they have on my later thinking about sentience.

First, the case of an accidental illusion. In 1985, a statue of the Virgin Mary was erected in a grotto close to the village of Ballinspittle in County Cork. Soon after the installation, people began to report that, when they came to pray to the statue after dusk, they saw it rocking from side to side. The 'moving Virgin' soon became a national sensation, drawing large crowds to witness what many took to be a divine miracle.

I went with a TV crew to report on it. On the evening we arrived in the village, we went to have a look for ourselves. The statue was dimly lit by a lamp from below, and around the head there was a halo of bright lights. As I stood staring at it in the dark, I was taken aback to see that the face of the statue seemed to be moving relative to the halo so that the whole statue rocked this way and that. To my surprise, I found myself seeing exactly what others had claimed.

We trained a video camera on the statue and filmed it. The video showed that the statue remained completely still. It wasn't until I and the team returned home and we had a chance to experiment that we worked out what was going on. On a computer screen we displayed a picture of the Virgin's face, dimly illuminated and crowned by the bright halo. We found that, if you stood and stared at this in the dark and the image was moved from side to side on the screen, the head did indeed appear to move within the halo.

Remembering my lectures on visual physiology, I now guessed the explanation. It's a known fact that receptor cells in the eye respond more quickly to bright lights than dim ones. This means that when the image of a bright object coupled to a dim one moves across the retina, the movement of the bright object will be detected first, with the result that the dim object will appear to lag behind. Now, the statue in the grotto did not actually move as our computer mock-up did. Nonetheless, anyone standing and staring at the statue in the dark would be bound to be slightly unsteady on their feet, and their eyes would move to compensate. These unconscious eye movements, in combination with the brightness-induced delay, would have been sufficient to create the 'miraculous' rocking that I and the worshippers observed.

The 'moving Virgin' illusion was unexpected: a serendipitous outcome that the creator of the statue certainly hadn't planned. But the next illusion we investigated was something very different.

*

In 1879, in the village of Knock in County Mayo, a group of villagers witnessed what they took to be a miraculous apparition. On the gable wall of the church, just after sunset, there appeared a ten-feet-high image of the Virgin Mary flanked by two Saints. The image remained for about two hours, growing brighter as night fell. The figures neither moved nor spoke. Fifteen witnesses testified to what they'd seen and described how they marvelled at it and discussed it among themselves.

News of the apparition spread. People came from far and wide to touch the church's wall. Soon, there were reports of prayers being granted and of miraculous cures. To cut to the present day, the site is now a major Christian shrine, Knock has an international airport, and the village priest, Archdeacon Cavanagh, is on the path to sainthood.

In this case, unlike the previous one, there seems to be no reason to doubt that what the witnesses saw was a genuine optical phenomenon—a real luminous image on the wall. However, from the beginning, there were many people who maintained that the image was produced by trickery. And the perpetrator of the trick was probably the priest himself.

Cavanagh, at the time, was in deep water politically. He had made himself highly unpopular with his parishioners by siding with British landlords, and there were threats to expel him. He desperately needed a public sign that he retained God's favour. Elsewhere in Europe, in the 1870s, miraculous apparitions had proved highly effective in restoring the church's authority. But such divine apparitions were not, of course, available to order.

So, it seems all too likely that Cavanagh decided to take matters into his own hands.

The question—then, and now—was how the trick could have been done. Given the descriptions of the image, it was widely rumoured that it was projected by a magic lantern. But where could the lantern have been hidden? And why did people approaching the wall not interrupt the beam?

Looking at the problem anew for the TV film, we noted that there was a small window high up in the gable wall. Perhaps, if the lantern was set up inside the church projecting outwards, its beam could have been reflected by a mirror down onto the wall.

We decided to experiment. There was an old granary with a similar window, next to my parents' house in a village near Cambridge. We borrowed a Victorian lantern and some religious slides. We set it up inside, on top of a ladder, projecting out of the window, and used a shaving mirror to reflect the beam down onto the outside wall. As night fell, the glowing image of Mary and her saints appeared—witnessed and admired by our incredulous neighbours.

*

I assumed there had to be a sharp difference between illusions of the supernatural that arise by accident and those that are contrived by human agency. Ballinspittle and Knock illustrated this perfectly. However, I saw later that there could be cases where it's actually a bit of both: first, an accidental illusion arouses people's wonderment, and then an enterprising fraudster—seeing the main chance —might step in to embellish it. Imagine, for example, if there were to be a statue like the one in Balinspittle that just happened to appear to move, and then—since everybody loves a moving virgin—the priest of the neighbouring parish were tempted to make *his* statue move for real.

Would it have to be fraud? It could be art. A magnificent example of art that was apparently inspired by accident may be seen in palaeolithic cave painting. It has often been remarked that the artists who decorated the walls of caves such as Lascaux and Chauvet incorporated existing features of the rock into their paintings of animals. It seems more than likely that the artist, looking at a virgin rock face, saw the head of a horse, the shoulder of a bison, the mane of a lion already prefigured there, and was duly amazed; then he or she carefully applied paint to fix and exaggerate the fleeting impression.

I made a note to myself going forward. Could the properties of phenomenal consciousness have arisen by some such happy accident, then to be taken up and embellished by natural selection as something like a work of art—just because, on a surreal level, they enrich their bearers' lives?

Aldous Huxley, in his novel, *Brave New World,* mockingly defined a philosopher as 'a man who dreams of fewer things than there are in heaven and earth'. But for humans, the shoe is on the other foot. Consciousness could be a dream of things that may exist nowhere at all but that could make life worth living nonetheless.

5

WHAT THE FROG'S EYE TELLS THE MONKEY'S BRAIN

I began research for a PhD in psychology in 1964. My supervisor now was Larry Weiskrantz, a man in marked contrast to Brindley, not so fiercely clever, but much kinder to himself and others.

Larry was born in New York to immigrants from Germany. When he was aged six, his father suddenly died, and his family lost their sole source of income. His mother had no choice but to send Larry to a free boarding school for 'poor male white orphans'. Within this school and, as Larry would say, 'only in America', he flourished, progressing by scholarships to Harvard, Oxford, and Cambridge.

In Cambridge he began a programme of research into the brain mechanisms underlying vision in monkeys. A big question was the role of the cerebral cortex in vision. In all mammals, there are two brain pathways for processing information from the eyes, an evolutionarily ancient one and a more modern one. The ancient pathway, which is also present in vertebrates such as fish and frogs whose brains do not have a cortex, runs from the eyes to the optic tectum in the midbrain. The other pathway, that evolved in the mammalian line, runs to the primary visual cortex.

Larry had been studying the effects of surgically removing the visual cortex from a monkey's brain. His research at this point had largely confirmed the conventional wisdom: that the operation left the monkey for all practical purposes completely blind. True, the monkey could still learn to choose between cards of different brightness and between a uniform grey card and a checkerboard pattern, but it was quite unable to discriminate the position or shape of objects. 'The simplest hypothesis concerning the capacity of this monkey is that it responded only to the integral of all retinal ganglionic activity. There is no suggestion it could respond to the distribution of variation.'[15] In other words, it seemed that monkeys' eyes were simply serving as light buckets without providing any information about the spatial pattern on the retina.

This was in line with earlier findings. Yet, a question hung over it. The monkey's midbrain visual system was still intact. Fish and frogs can see just fine using the optic tectum. Why should the monkey become so incapacitated visually after the operation?

*

It was agreed that I should embark on a study of single nerve cells in the monkey's optic tectum (also known as the superior colliculus) to see just what kinds of visual information this secondary system might be capable of processing. Since no one else in our lab knew the techniques of recording from single cells, Larry sent me to Edinburgh for a couple of months to be shown the ropes by the renowned neuroscientist David Whitteridge.

Whitteridge took me under his wing. He taught me how to make fine-tipped needle electrodes for recording the electrical activity of cells in the brain. Then, he demonstrated how to surgically open up a hole in the skull of an anaesthetized cat and insert the needle through the brain to just the right position in the visual system close to a responsive nerve cell. When the cell fired, the discharge was picked up by the electrode and amplified so that we could hear it on a loudspeaker.

Back in Cambridge, I adapted these techniques to recording from cells in the superior colliculus in monkeys. The anaesthetized monkey was positioned in front of a screen on which I could move small black or luminous targets. I arranged that the position of the target was displayed as a spot on an oscilloscope and linked the brightness of the spot to the cell's response so that it lit up only when a spike occurred. This meant a picture would emerge on the oscilloscope of the cell's receptive field—that's to say the area of space that fell within the cell's 'field of view'.

My job was to find what kind of visual stimulus would excite the cell I was recording from. It turned out that cells responded best to moving targets, crossing the screen in any direction, at about 10 degrees a second. Cells near the surface of the colliculus had very small receptive fields, meaning they could potentially pinpoint exactly where the target was situated; but as the electrode went deeper the fields became much larger, meaning they were indifferent to the target's exact location.[16] Figure 5.1 shows results for a range of different cells.

These were new and interesting findings. They showed that the surface layers of the colliculus could be relaying information to the rest of the brain about a target's position and so might in principle be able to support spatial vision—even if this didn't fit with the behaviour of Weiskrantz's operated monkeys. This would be an important lead to follow up. But while I was doing the experiments, I confess the nice 'results' I was getting were not

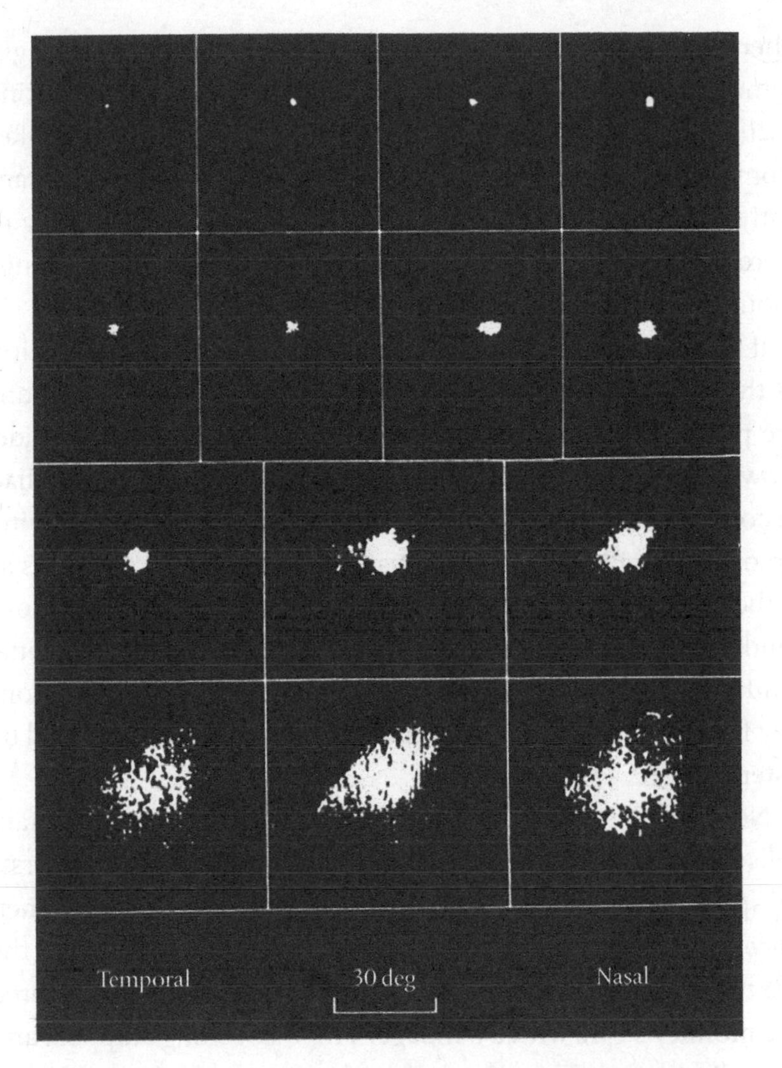

Figure 5.1 Receptive fields of cells in the superior colliculus of the monkey in response to a moving 1.5-degree black disc or 1-degree luminous disc[16]

always what most concerned me. I look back now and my heart misses a beat.

I was a twenty-three-year-old student working alone, often deep into the night, in a blacked-out room in a deserted building.

There was an anaesthetized monkey tied to a chair. The only light came from the targets moving across the screen and the winking oscilloscope; the only sound was the sporadic crackle of spikes from the loudspeaker. The animal would be put to sleep permanently when I was done with it. The brain cells I was listening to were 'seeing' for the last time. In this borderline situation, strange thoughts swam through my mind.

If the animal were awake, it would be having a visual sensation as the stimulus moved across its retina. But because sensations are private, no one would be able to tell this from the outside. Now, however, some part at least of this private experience had become externalized: the animal's response to light was showing up on the oscilloscope and activating the loudspeaker. It was as if the animal was expressing how it felt about the stimulus out loud—growling, say, or purring as the light stroked its retina. And here was I, listening in. But if I could listen to how the monkey felt about the visual stimulus, perhaps the monkey could be listening too?

Next, poetry took over. I had already come across certain cells that responded to a moving target with a series of bursts of spikes rather than a regular string: *whoosh*! a pause, then *whoosh*! (See Figure 5.2.) What was this about? On a hunch, the next time I found a cell responding in this odd way, I covered the monkey's ears with a bandage. The whooshing stopped, and the cell responded to the target with a steady discharge. Then I uncovered the ears and turned down the volume of the speaker so that I could only just hear it. Again, no whooshing.

This must have been what's called a 'multi-modal' cell—a cell that had input from *the ears as well as the eyes*. I had found other examples, although they were relatively rare. Could it be that, when the volume of the speaker was turned up, the cell was first responding to the visual target with a burst of spikes, then

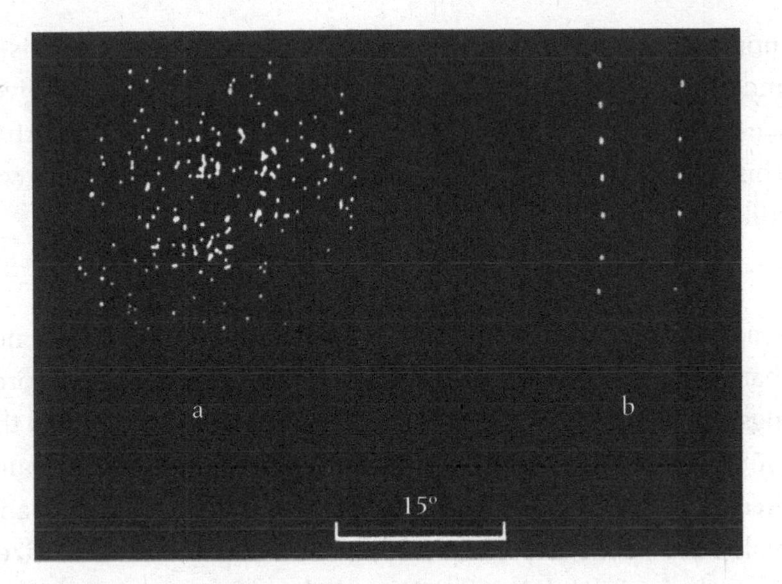

Figure 5.2 Receptive field of a cell in the monkey's superior colliculus, in response to 0.5-degree luminous disc. In (a), the field has been crossed many times, in (b), twice only, at a speed of about 10 deg/sec. Here, each dot corresponds to a high-frequency burst rather than a single spike; note the discontinuous response[16]

responding to the sound these gave rise to with more spikes, and then to the sound *these* gave rise to? The result would be positive feedback, creating a brief surge in activity that would exhaust itself. Whoo...sh! To test this, without a visual target, I clapped my hands. Sure enough, the cell responded with a whoosh to the sound of the clap. The effect of the auditory feedback was that the cell's response to any stimulus was getting drawn out in time, producing a kind of afterglow.

The set-up in the lab was, of course, completely artificial. But this unexpected phenomenon gave me the seed of an idea. We tend to think of sensations as experiences that are impressed on us from the outside. But suppose our sensations actually originate as an active bodily response to the stimulus, like the signals sent to the speaker, and that we only become aware of this when we monitor our own response by a kind of listening in? Could this—and the feedback loops that would easily follow—be what gives sensations the thick, expressive quality that we find so wonderful? This idea would later take root in a theory of qualia. But that was still to come.

*

I may say, I didn't much like doing these experiments. It's not that I doubted their scientific value. These were the first recordings ever made from the superior colliculus of monkeys (and the 1968 paper where I described them has been cited several hundred times). Nor was it that I thought experiments on living animals were wrong in principle. The monkeys were anaesthetized throughout and didn't suffer. Nonetheless, there was no denying that what I was doing had a worrying power dimension. It could have been said (no one did, but I thought it) that—to put it bluntly—I was valuing my curiosity about how the monkey's brain works over the monkey's interest in enjoying the use of its brain. Of course, I hoped my findings would contribute to the larger project of understanding the neuropsychology of vision in monkeys and in humans, so at least it wouldn't be *idle* curiosity. But would they contribute?

The thing that certainly looked promising was the evidence that the cells I was studying were quite capable of relaying information about the location of objects in space. In this respect, the monkey's superior colliculus clearly did indeed resemble the frog's optic tectum. In fact, the kinds of stimuli the cells responded to were markedly similar to those described in a famous paper by Jerry Lettvin and colleagues in 1959, entitled 'What the Frog's Eye Tells the Frog's Brain'.[17]

After removal of the visual cortex in a monkey, this relatively primitive visual pathway could still be operational. In which case, the monkey might retain at least a frog-like capacity for spatial vision. Yet, Weiskrantz's research seemed to prove not. I began to wonder. Could Weiskrantz have been missing something? Maybe something that was hiding, not in plain sight, but in *some other kind of sight entirely*?

6

BLINDSIGHT

It was the summer of 1966. There was a monkey in the Cambridge lab from whom Weiskrantz had removed the visual cortex a year and a half earlier. Her name was Helen. After the operation, she apparently gave up on the use of her eyes. When left to herself, she would stare vacantly into the distance, never bothering to look around. And when tested, the best she could do was to discriminate between cards of different brightness. She was as good as blind.

That autumn, Weiskrantz went to a conference in Basel in Switzerland. I took my chance. Over several days, I sat by Helen's cage and played with her, hoping to get some hint that she was not quite as blind as everyone was saying—and as maybe she believed as well.

I tried to attract Helen's attention by talking to her, waving my hand and clicking my fingers. To begin with, she seemed reluctant to engage. Yet, I soon got the impression that, despite herself, she was sometimes following what I was doing. At least, she seemed to be looking in the right direction. To encourage her, I made this more worthwhile by holding a piece of apple in my hand and letting her take it if she could. She quickly caught on. Before long, I had her reaching out to touch a black-and-white cube that I twiddled on the end of a stick, and then a flashing pea-bulb, and then a stationary lighted bulb, and then a black-and-white cube that didn't move.

I sent Weiskrantz a telegram to the Basel hotel: LARRY. STOP. YOU WONT BELIEVE IT. I'VE TAUGHT HELEN TO SEE. STOP.

He didn't believe it. When he returned to Cambridge, he was at first too busy to find time to see what Helen and I had been up to. Of course, no professor takes kindly to having a PhD student undermine results he thought were in the bag. However, when he got around to seeing it for himself, he was immediately won over and was as keen as me to see where it would lead. We jointly submitted a paper to the journal *Nature* describing how a monkey with no visual cortex could in fact detect the spatial location of a visually salient object.[18]

But we had jumped the gun. As it turned out, Helen's vision had a lot further to go.

*

Helen was scheduled to be killed and her brain examined to confirm the extent of the surgery. But Weiskrantz agreed that I could go on working with her. I took her with me to the Oxford Psychology Laboratory when I moved there in 1967, and then back to Cambridge in 1971 to the Department of Animal Behaviour in the village of Madingley. For seven years, I was her tutor almost on a daily basis. I encouraged and coaxed her, trying in every way to help her realize what she was capable of.

In the Oxford and Cambridge psychology labs, I always tested her in the small cage where she was housed. But when we came to Madingley, I found myself among scientists with a different outlook. In a typical psychology lab, monkeys were studied to gain insights into how the minds of humans work. At Madingley, monkeys were studied to find out how the minds of monkeys work. The emphasis here was on whole animals in their natural environment and on animals as individuals.

A tell-tale sign of my own background was the language I used in print. In the *Nature* paper, I didn't call her Helen but abbreviated it to Hln. More telling still, I didn't call her *she* but *he.* In later papers, I did give Helen her name and sex back. Even so, it never occurred to me to mention a fact of some importance to how she performed in tests. This was that every thirty days or so her mood changed and she became quite 'difficult': she was having her menstrual period. Too much information.

It so happened that a young French primatologist, Mireille Bertrand, was visiting Madingley. Mireille, who studied wild monkeys in Africa, was upset to see Helen in her cage. 'Why don't we try letting let her out', she asked, 'so she *has* to use her eyes?' 'But she won't manage', I said. 'She's been in that cage for years, she's never been out and she's not tame.' 'Then I must tame her.'

I watched through a glass door as Mireille went into the room where Helen's cage was mounted on a rack. She undid the door of the cage and left it open while she stood to one side. Helen hesitated at the threshold. She gingerly climbed out, let go of the cage wall, and dropped to the floor. Then she panicked. She leapt in the air and landed on Mireille. Mireille grabbed her arms. Helen tried to bite her. Mireille bit Helen back. The two of them wrestled. Finally, Mireille dominated her and pinned her to the ground. And then, of a sudden, there was peace between them.

Next day, Mireille put a dog collar on her. We attached a lead and took her for a walk outside the building. Helen was by no means completely tame, but she adapted quickly to the lead and soon we were able to explore the meadow and woods around the laboratory. To begin with, as might be expected, these walks were fairly hazardous. Helen continually bumped into obstacles; she collided with my legs and several times fell into a pond. But

Figure 6.1 Helen at large

over the next few weeks, things markedly improved. She soon began to anticipate and skirt round obstacles in her path. Still more remarkably, she would single out a tree in the field, head over to it, and climb the trunk.

There was an old elm tree that she specially liked to climb. So, with her perched in a hole in its trunk, I would hold up bits of fruit and nuts for her to reach to. And now she did something else I never expected. She would reach out when the target was within arm's length *but ignore it if it was too far away*. It was clear that, at last given the experience of three-dimensional space, she was not only developing three-dimensional visual perception but perception she could monitor introspectively—she *knew* when something was out of reach.

*

I set up a big indoor arena with moveable furniture. Soon, Helen was completely at home in it. She would run around, avoiding baffles and obstacles, while finding and picking up small currants from the floor. Her vision soon proved to be so acute that it was hard to keep the floor clean enough to prevent her trying to pick up specks of dirt. When twenty-five currants were scattered at random over an area of 5 square metres, she took less than a minute to find every one. To anyone who was unaware of her history, it would have seemed that she was seeing quite normally.[19]

Yet, I was becoming convinced that Helen was *not* seeing normally. I knew her too well. We had spent so many hours together, while I continually wondered what it was like to be her. I found it hard to put my finger on what was wrong. But I had the hunch that, despite the evidence, she still did not *expect* to be able to see. She seemed strangely unsure of herself. If she was upset or frightened, for example, her confidence would desert her and she would stumble about as if she was in the dark again. When the doyen of neuropsychology, Hans Lukas Teuber, made a visit from the Massachusetts Institute of Technology especially to observe Helen's visual recovery, she embarrassed us both. His presence made her too nervous, and while he was in the room she acted blind. It was as if Helen could only use her vision when she was relaxed enough not to think about it.

In 1972, I wrote an article for the *New Scientist*, and on the front cover of the magazine they put the headline, under Helen's portrait, 'a blind monkey that sees everything'. But this headline wasn't accurate. Not *everything*. My own title for the article inside the magazine was 'Seeing and Nothingness', and I went on to argue that this was a strange kind of seeing that might be stranger than we could imagine. I wrote: 'When people suffer extensive damage to the visual cortex it is said that their blindness is total and permanent. Perhaps with a more flexible definition of vision, it will yet be discovered that there is more to seeing than has so far met either the clinician's or the patient's eye.'[20]

With a monkey, who could not describe her inner world, there seemed no way of knowing what her experience was really like. To find out, we would need evidence from human beings, and at that time there were no human cases that resembled hers. Indeed what evidence there was—including a new study by Brindley—suggested that humans with similar brain damage were permanently blind.

*

Two years later, Weiskrantz made a dramatic discovery. He was studying a human patient known as D.B. who had undergone a brain operation at a London hospital to cure intractable headaches. The operation involved removing the visual cortex on the right-hand side of the brain. This resulted in the patient's immediately losing all vision in the left half of the visual field. When a light was presented in that part of the field, he denied that he could see it. However Weiskrantz, emboldened by the discoveries with Helen, gently leaned on him to try something that seemed impossible. He asked D.B. to point to where he *guessed* the light might be. And to everyone's astonishment, not least the patient's, he consistently got it right. Further tests showed he could guess not only the position of an object but also its shape and colour. Yet, all along he insisted he was unaware of any visual sensations.

Weiskrantz named this capacity for insentient vision 'blindsight'.[21] In 1974, he sent me an offprint of the paper where he described the first findings with D.B. On it he wrote HELEN IS VINDICATED. I believed it.

But again, we were jumping the gun. Blindsight in humans had a lot further to go. In fact, in several respects it was going to go further than with Helen. Subsequent studies have revealed that humans with damage to the visual cortex can, in the 'blind' areas of their visual fields, assess the three-dimensional shape of an object before they reach to it, recognize emotional expressions in faces, and maybe even read and understand written words.

None of the patients first studied reached the same level of skill as Helen at visually guided navigation in space. But then, in 2008, a patient who had no visual cortex at all after suffering repeated strokes and who considered himself totally blind proved to be able to walk down a cluttered hospital corridor avoiding every obstacle.[22] There's a film of him doing this. It shows Larry Weiskrantz in the background, at age eighty-two, with a broad grin on his face.[23]

*

From early on, I wondered how Helen and the phenomenon of blindsight might be able to throw light on the evolution of sentience in humans and other animals. In the absence of a visual cortex, Helen's vision had to be mediated by the superior colliculus, the evolutionary descendant of the frog's optic tectum. But does that mean she was actually seeing like a frog? Or, to put this the other way round, do frogs actually see like Helen? Do frogs have blindsight? If so, is the same true for all those other vertebrate animals whose brains have not evolved a secondary visual system and whose vision is mediated by the optic tectum—that's to say, all but mammals and birds? Would it be true also of human infants, in whom the cortical visual pathway does not mature for several months after birth? How surprising—but how interesting—if vision without any accompanying conscious sensation is the way we all start out and remains that way in many animals today.

On another level, I wondered how to characterize blindsight philosophically. I found myself falling back on Thomas Reid's crucial distinction between sensation and perception. A patient with blindsight manifestly has a form of visual perception, an ability to detect the properties of external objects in the blind part of the visual field. But he has none of the usual sensations that would normally tell him about the light at his eye. As far as he's concerned, he's merely 'guessing'. A 'guess' is defined in the dictionary as 'a judgment or opinion without sufficient evidence or grounds'. Exactly so. The patient is no longer aware of having sufficient grounds for being able to perceive: it seems it has nothing to do with *him*. Blindsight is a case of *pure perception in the absence of sensation.*[24]

I read further and was impressed to find that Reid himself had argued in his 1764 book that sensation and perception could be dissociated. 'We might perhaps have been made of such a constitution, as to have our present perceptions immediately connected with the impressions upon our organs of sense, without any intervention of sensations.'[25] And in a letter of 1778, 'I can conceive a Being that has Sensations of various kinds without any Perception . . . I can conceive also a Being that perceives all that we perceive without any Sensation connected with those Perceptions.'[26]

So, Reid could conceive of something like blindsight! Yet, he would have been the first to agree that it would seem to be contrary to common sense. For, as he wrote, '[under normal circumstances] the perception and its corresponding sensation are produced at the same time. In our experience we never find them disjoined. Hence, we are led to consider them as one thing, to give them one name, and to confound their different attributes.'[27]

Given this constant conjunction, common sense would suggest that there has to be a causal relationship. Presumably sensation, being more elementary, comes first and lays the ground for perception. But, in that case, blindsight— perception without sensation—would not just be unusual, it would be logically impossible. The existence of blindsight shows how we typically misapprehend how our own minds work. It seems we are taken in by a 'user-illusion' that, by giving our sensations a causal role, provides us with a story about *how* we are able to perceive. Blindsight would seem to be selfless sight, which—psychologically—doesn't seem to make much sense.

However, while we end up getting this wrong, it's worth noting that the reality about how the brain handles perception would not surprise an engineer. If an engineer were designing a robot so that it could use its camera eyes to explore the external world, he or she would never do it in two stages: first, come up with a description of the image at the camera and then use this description as the starting point from which to deduce what's out there. They simply wouldn't bother with describing the image as such. Instead, they'd use filters to separate out different categories of information—about position, movement, shape, colour, and so on—and then combine these components algorithmically so as to supply what the robot needs to know about the world.

As it happens, the famous paper 'What the Frog's Eye Tells the Frog's Brain' was published in an engineering journal, the *Proceedings of the Institute of Radio Engineers*. And in it the authors explicitly invoke the sensation/perception distinction. They remark that the kind of detectors they discovered in the frog's visual system were apparently purpose-built for *perceptual tasks* such as 'bug detection'. 'The operations thus have much more the flavor of perception than of sensation…That is to say that the language in which they are best described is the language of complex abstractions from the visual image.'[28] It seems these pioneers of neuroscience would have had no trouble with the idea that frogs have blindsight.

Then, here's the question. If blindsight really is the natural condition of a frog, and presumably the condition of all man-made robots, we have to ask what would be missing in either animals or robots if they were to lack phenomenal consciousness completely. What would be wrong—or insufficient for survival—with deaf hearing, scentless smell, feelingless touch, or even painless pain? The killjoy hypothesis would be that *percipient but insentient* animals could manage perfectly well.

For me, this was to become the hypothesis that a theory of sentience would have to disprove.

7

SIGHT UNSEEN

There were many times when I tried to see things from Helen's point of view. One thing I took for granted was that my research with her had, at some level, made her life more worth living. Her routine had certainly become more varied and interesting. She seemed to positively enjoy being tested in the lab. At whatever level monkeys feel these things, I assumed she must be happy to find that she could use her eyes again. *I* felt good about it. Surely, *she* must too. It never occurred to me that, if Helen were a human being, she might find blindsight worse than no sight.

Then, in 1973, I became involved in the study of a young Iranian woman, known as H.D., which threw unexpected and tragic light on just this question. Dennett, in his review of *Seeing Red*, picked this out as one of the 'profoundly unusual encounters with the ill-understood boundaries of sight and experience' that he thought might have led me down a cul de sac. Since I've nothing to add to the account I gave previously, I'll tell it in much the same words.

H.D. at age three contracted smallpox, which caused severe scarring of the corneas of both her eyes and rendered her effectively blind. As we know from a memoir she wrote later, as a child she was exploited by her family, forced to beg on the streets, and sexually abused. But, against all odds, at age fifteen she was rescued by a Christian mission and sent to a school for the blind, where she learned to read braille and became fluent in speaking English. Six years later, she enrolled as a student at Teheran University. There, she was discovered by a London-based eye-surgeon who believed that, if she could be given corneal grafts, there was a good chance that she could recover normal vision. Her community raised the money to bring her to Moorfields Eye Hospital for the operation.

In 1972, aged twenty-seven, she arrived in London with high hopes. The operation went ahead and was technically a success. The optics of her eyes were restored. However, immediately after the operation there was no evidence that her vision had improved, and she continued to make crude random eye movements, as is typical of patients blind from early life. Two months later, things had got no better, and she was transferred to the Psychology Department of the National Hospital for assessment by the neuropsychologist, Elizabeth Warrington. I was invited by Dr Warrington to join her.

When I first met H.D, I found her in a state of despair. She was convinced the operation had been a complete failure. And no wonder: for, as far as she was concerned, she was as blind as before. Warrington and I realized that there was an all too probable explanation. When the visual cortex of the brain is not exercised by getting input from the eyes, degenerative changes are likely to occur. Because H.D.'s visual cortex had remained unused since early childhood, there was a real possibility that it was no longer able to function properly. In that case, she might now be in much the same condition as my monkey Helen, whose visual cortex had been surgically destroyed. Unfortunately, new eyes could not compensate where a new brain was needed.

But Helen, when I first met her, was also convinced she could not see. If H.D.'s case was in some ways like Helen's, then perhaps she might be capable of learning to see again, as Helen had.

At any rate, I thought it worth trying some of the same methods that had worked with Helen. I took H.D. out to 'see' the sights of London. We walked the streets and parks while I held her hand and described what was in front of her. And, to our delight, it soon became clear—to her as well as to me—that I wasn't leading a blind person. *Something* had changed from before the operation—something that could even be of practical value. She could point to a pigeon as it lighted on the square, she could reach for a flower, she could step up when she came to a curb. She was indeed recovering an ability to use her eyes to guide herself through space.

Thus it seemed that the operation had not been a total failure after all. H.D.'s eyes and brain were working together again. However, she herself remained far from happy with her progress. Although she had recovered her sight to some degree, it was in no way the degree she had been hoping for. In fact, it only left her feeling sadder and more cut off. For the awful truth, she let on, was that—just as in blindsight (and maybe it really was a kind of blindsight)—her vision lacked any subjective sensory quality. She had been living for twenty years with the idea of how marvellous it would be if only she could see like other people. She had heard so many accounts, stories, poetry, about the wonders of visual experience. Yet now, here she was with part of her dream come true, and she simply could not feel the benefits.

Coming to this from thinking about *what it would be like* to have blindsight, I guessed that what made H.D.'s vision so relatively worthless was that, lacking phenomenal quality, she didn't experience it as *hers*: indeed, it didn't contribute to her sense of *self*. She felt cheated. It was a mockery of what she had imagined. In the event, she soon refused to go on being tested in the clinic. She no longer wanted to go out to be shown the sights by me. In fact, she began to resent it. As we wrote in our scientific research report, '"Seeing", far from being a rewarding activity, had become a tiresome duty for her, and left to herself she soon lost interest in it.'[29] She became increasingly depressed, almost suicidal. Finally, with great courage, she took back control of her situation—by putting on her dark glasses again, taking up her white cane, and going back to her former status of being conventionally blind.

This was certainly an exceptional case, and I would agree with Dennett that there's a danger of reading too much into it. Nonetheless, H.D.'s case stayed with me—to remind me, if I should ever forget, how much phenomenal consciousness can matter to a person's sense of who they are. To recall the challenge at the end of Chapter 6, here was a living example of someone for whom percipience without sentience was not enough. H.D. could not 'manage perfectly well' with the vision of a frog.[30]

8

RED SKY AT NIGHT

In 1967, I moved with Weiskrantz to the Institute of Psychology in Oxford, a Victorian villa on Parks Road. I continued to work with Helen. But in a hut in the bottom of the garden, I also started on a new line of research. Ostensibly, this was about animal aesthetics: I was going to test whether rhesus monkeys had likes and dislikes for particular colours. For me, it would be a change in direction away from studying brains. But I also had a hidden reason, which I didn't let on. I wanted to approach the problem of consciousness from another angle.

The set-up I used to test for preferences between colours was quite simple. I had the monkey sit in a small dark chamber, of which the back wall was a translucent screen which could be illuminated by a projector. When the monkey pressed and held down a button in front of him, light of one or the other colour flooded the screen. When he let go, the light went out and he was in the dark. Then, when he pressed and held again, he got light of the alternative colour. Monkeys don't like to sit in the dark, so typically they would keep the button pressed most of the time; but, it seemed, they also like change, so every so often they would let go and switch between the colours. By adding up the total time they stayed with each colour, respectively, I hoped to get a measure of which they preferred.

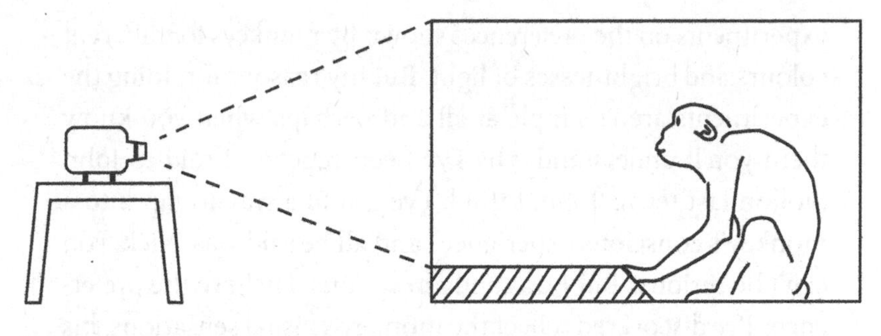

Figure 8.1

The results were dramatic. Every one of ten monkeys showed the same strong pattern of preference. They spent most time with light at the blue end of the spectrum and least with light at the red end (with the brightness made equal). Given a straight choice between blue and red, they would spend more than three times as long with blue as with red.[31]

*

It would be two years before I somewhat shyly explained to my colleagues what I was up to. I've recently come across the text of the talk I gave at a departmental seminar in 1969. Here's how it started:

> When people in the Institute have asked me what I've been doing in the last year or so, down there in that hut, I may have replied rather blankly 'Oh, experimental aesthetics in monkeys—you know, the appreciation of colour and so on.' And whoever I've been talking to has given a polite but questioning smile. They may perhaps have said 'How interesting. Do you know it's funny how Conrad always picks out something green . . .' And there the discussion has come to an awkward end. I realise I'm at fault, because I haven't properly explained to anyone what I'm about. Today I want to try and come clean. I shall be describing some rather simple

> experiments on the preferences shown by monkeys for different colours and brightnesses of light. But my reasons for doing the experiments aren't simple at all and perhaps when you know them you'll understand why I've been reticent. I said to John Mollon last term: 'John, I think I've found a way to tap into a monkey's conscious experience', and all he said was 'Nick, you can't be serious.' But I was and am serious. I believe the preferences I've discovered reflect the monkey's visual sensations, his subjective feelings about the colour of light at his eyes, rather than his perception of coloured things out there in the world. What we're seeing is the qualities of conscious sensation, experienced in private, coming out in public.

It was, I suppose, a rash claim. The neuroscientist, Richard Passingham, remembering the atmosphere in Oxford in those days, has written: 'When I was a student, mention of consciousness would have sent my tutors into convulsions. Only with Larry and blindsight did the study of consciousness become respectable.'[32] But this seminar was before human blindsight had been discovered. And so far I'd kept my thoughts about the nature of Helen's vision to myself.

To make my case in the talk, I used a roundabout argument, harking back to Reid. It went as follows. Sensations are mental states that concern what's happening at the subject's sense organs. Perceptions concern the existence of objects in the external world. Suppose we want to study the psychological effects of sensations on their own, independent of perception. Then, we might be able to do this if we set things up so that two very different sets of stimuli at the body surface reveal the existence of one and the same thing in the external world. If the subject evaluates these stimuli differently, this must be attributable to sensation, not perception.

So let's think about an experiment involving 'sensory substitution', with you as the subject. (It could only be a thought experiment because in 1969 no one had yet tried anything like this in practice.) Consider what it would be like for you if the information about sounds in the environment, normally processed by your ears, were to be made available instead to your eyes in the form of a sonogram on a tv screen. We'll assume, for the sake of argument, that after enough practice, you'd be able to perceive visually all the things you normally perceive aurally. So, given that your perception of external events would be the same, would it *matter* to you which sensory channel was in use?

If you were a sound-perceiving robot, I think the answer might be: no, you'd be indifferent to the modality of the stimulation. But given you're a human being, the answer must be: yes, it obviously would matter—not necessarily always, but certainly sometimes. Just imagine, for example, the difference between hearing a baby's cry with your ears and seeing it displayed on a screen with your eyes. Or the screech of chalk on a blackboard... or a piano sonata.

Wouldn't this be a great experiment to do with monkeys! If the monkeys were to show that they preferred one kind of sensory stimulation to another, even when the perceptual information stayed constant, it would surely be prima facie evidence that their choices were—as with humans—being governed by the quality of their sensations.

Regrettably, I had to admit this wasn't the experiment I'd actually been doing. Sensory substitution with monkeys just wasn't a feasible proposition. So, I had settled for what was arguably the next best thing. Rather than giving the monkeys a choice between different stimuli that carried the *same* perceptual information, I'd given them a choice between stimuli that in effect carried *no* information about the external world.

In the testing chamber, there was essentially *nothing to look at.* The monkey's eyes were flooded with coloured light and there were no *coloured objects* of interest. What's more, the change in colour of the screen from one button press to the next was completely predictable. So, I reasoned, if the monkey in this situation did, in fact, prefer to sit in blue light rather than red, this could not be because he liked *blue things better than red ones*; it would have to be because he liked *the sensation of blue light reaching his eyes better than the sensation of red light.*

Having explained the rationale, I proceeded to describe the results of my experiments so far. And I ended the talk with a flourish:

> The fact that I've found these behavioural preferences proves that the monkeys have strong subjective feelings about the colour of light reaching their eyes. These feelings must surely reflect the qualities of the sensations they are experiencing. I hope you'll agree this justifies the claim I made at the start—the one that drew John Mollon's scorn—that I've found a way of outing the monkeys' conscious experience.

Of course, not everyone agreed. I gathered afterwards that a senior colleague, Jeffrey Gray, let it be known that he could drive a coach and horses through my presentation. However, he was kind enough not to tell me this directly, so I never heard his reasons.

*

Looking back, I can see that the philosophical argument had gaps. I'll address some of the problems later. But if there was something I'm embarrassed to have got wrong in that seminar, it's not the philosophy, it's the science. For, as I soon came to realize, there was an alternative way of interpreting the experimental results that I'd been getting.

I had set out to test for colour preferences, and preferences were what I believed I found. As I said, the monkeys used the button to alternate between the colours. And on average they did indeed spend much longer with the button held down for blue light than for red. I interpreted this to mean they *liked* the blue light better. But was that right?

Rather late in the day, I remembered a classic experiment with woodlice. If a woodlouse is put in a box with a damp end and a dry end, it will wander around apparently at random, but overall will spend much longer in the damp end. This is because *it walks faster* when it senses dryness in the air, and so will *probably exit* the dry end sooner. However there's no reason to think the woodlouse has any subjective preference: that it *likes* being damp better than being dry.

With my monkeys now, suppose they were to press and let go of the button at random but *were generally to do things faster in red light than in blue,* then they would end up spending longer in blue because they'd tend to let go sooner in red. But again, this might have nothing to do with subjective liking.

Therefore I realized I had to do the obvious control experiment. Instead of having the colour of the light change when the monkey let go of the button and pressed again, I should simply have it stay the same. So, for example, instead of red–blue–red–blue, it would go red–red–red–red or blue–blue–blue–blue. The question was: would the monkey keep letting go and pressing again for the *same colour* but tend to *let go faster when the light was red?*

When I ran the experiment, there was no doubt about it. Yes, the monkeys merrily continued to switch the light on and off even when the colour didn't change, but the hold-times were indeed much shorter in red than blue. I'd proved my earlier interpretation wrong. This had to do with *timing*, not liking.

*

I can't deny I'd been proud of the experimental set-up. Even if I now had to interpret the findings differently, I could say at least they were *real findings* about how monkeys are affected by coloured light. However, it soon dawned on me there might be another problem. Real need not mean realistic or natural. Indeed I had to admit that, from the monkeys' point of view, the set-up was extremely artificial. Nowhere in nature do monkeys get to switch the colour of the sky, instantly, by pressing a button. At that point of my career, I hadn't yet come across the term 'ecological validity', meaning consonance with conditions in the wild, but clearly this wasn't it.

When I returned to Cambridge, to the department of Animal Behaviour in Madingley, I decided to re-examine the new findings using a set-up that came at least a little closer to a situation a monkey might encounter naturally. Instead of having things all happen in the same space, I built a box with two separate chambers connected by a short tunnel through which the monkey could pass as he wanted. The chambers were continuously lit by coloured light projected onto a screen at each end.[33]

Figure 8.2

It turned out that the monkeys' behaviour in this situation duplicated exactly what I'd found previously. When one chamber was blue and the other red, they would sit for a while in the blue, then get up and go and sit in the red, and then keep shuttling to and fro but waiting on average longer in blue than in red; and if both chambers were blue or both were red, they would still keep moving but again generally waiting longer in blue than in red.

Watching on closed-circuit television, I found their behaviour intriguing and not easily explicable. There seemed to be nothing stereotyped or mechanical in the way they moved between the chambers, nor were they wandering about aimlessly and accidentally entering the tunnel; rather, every move looked purposeful. The monkey would be sitting apparently contentedly, then suddenly he would become alert, glance around, and take off quickly through the tunnel to the other side.

However, when I analysed the detailed timing of their decisions to move, I found to my surprise that here *was* something mechanical, almost clock-work, going on. I measured the 'bout lengths'—how long the monkey stayed before moving. These varied from a couple of seconds to thirty or more. From these data, I calculated 'survivorship graphs', showing the likelihood that the monkey would sit at least so long before moving.

Figure 8.3 shows the graphs, averaged over seven monkeys, when both sides were blue or both red. If you look at the percentage of bouts that lasted at least ten seconds, you can see that in blue light it's about 50 per cent and in red light 30 per cent; for bouts that lasted thirty seconds, in blue light it's about 10 per cent and in red light just 3 per cent. (Note that the y axis has a logarithmic scale.)

What's truly remarkable is that the graphs for both colours are straight lines. This means that the bout lengths conform to a 'Poisson distribution', where the probability the monkey will

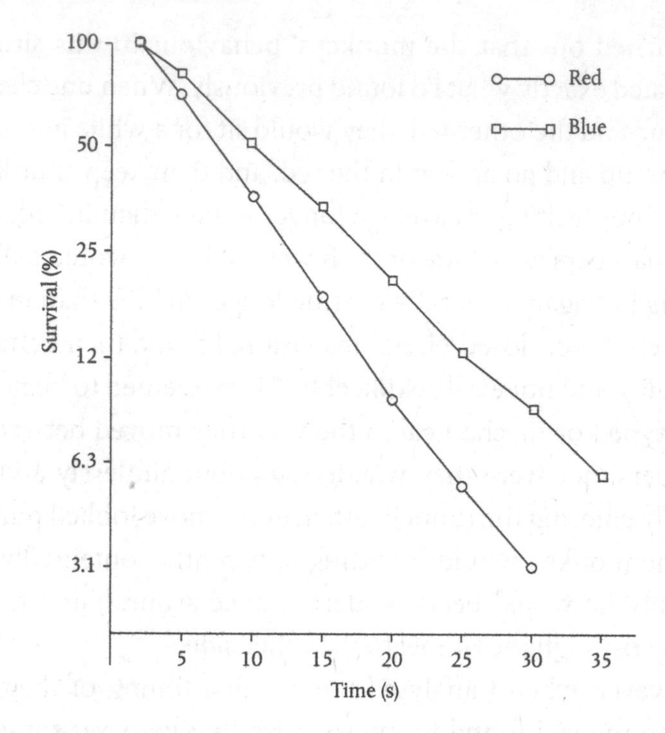

Figure 8.3 Log survivorship curves for bouts in red and blue light[33]

decide to move in the immediate future remains the same no matter how long he's already been sitting.

To picture this in terms of a simple model, suppose that every *H* seconds the monkey tosses a coin: if the coin comes down heads, he moves; if it comes down tails, he stays where he is—and *H* seconds later tosses the coin again. Then, sometimes he will get a heads and move after the first toss, but sometimes he will wait for several tosses before doing so. The shorter *H* is (that's to say, the more frequently he tosses), the sooner he is likely to get a heads and move.

H corresponds to the slope of the graph. As you'll see, the slope was greater in red light than in blue. This is just what would happen if the decision whether or not to move was coming round more rapidly in red than in blue—in fact, about 50 per cent more rapidly.

*

Here was striking confirmation that colour was affecting timing rather than liking. But, what's more, here was evidence that this was happening in a surprisingly routine way. It may not have been the result I was originally looking for, but it served the goal of bringing the monkey's private experience out into the open even better. Whoever would have guessed that the quality of visual sensations would have such a straightforward influence on a simple behavioural variable as decision frequency?

The new experiment also threw light on something that had been a puzzle up to that point. Every monkey I had ever tested, thirty or so, independent of their sex and age, showed the same response to colours. It seemed clear that this had to be an evolved trait, hard-wired into the brains of monkeys by natural selection. This meant that it must have survival value for monkeys in the wild. But because my original testing situation was so unnatural, it had been hard to imagine just what this value might be.

The new situation, however, should be easier to relate to the behavioural ecology of monkeys in nature. The first thing to explain was the monkeys' *shiftiness*. Why ever did they go on shuttling between the chambers, apparently purposefully, even when the chambers were identical? To me, who *knew* there was nothing to be learned by moving, it seemed a complete waste of effort. Yet, did the monkey know what I did? How could he be sure?

Look, for yourself, at the Necker Cube in Figure 8.4. You see it first this way round, then that way, then this way again. You ought to know nothing is going to change. Yet, it's as though a kind of mental tic drives you to go on 'sampling' each of the

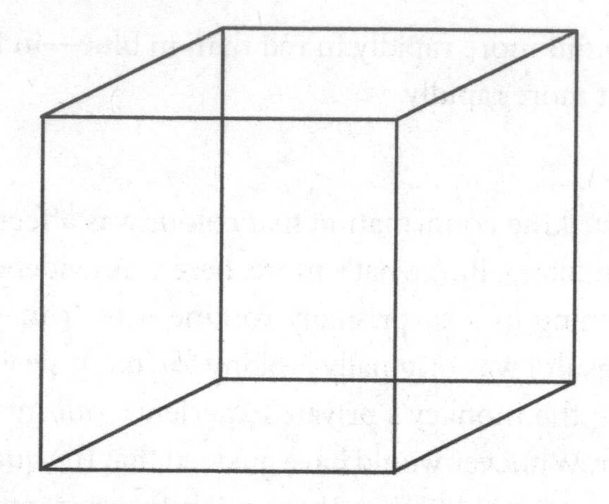

Figure 8.4 The Necker cube

possible alternatives, just in case. I imagine it was likewise with the monkeys. Being able to be in only one place at one time, they could never be sure they were not missing something in the other chamber—and every so often the urge came over them to check on it.

In the real world, such periodic checks would clearly be adaptive. The real world, unlike the testing box, cannot be counted on to remain stable over time. If monkeys are to keep themselves informed, they must regularly sample hidden bits of the environment. But they mustn't be either too obsessive or too nonchalant about it. The sensible strategy will be to space out successive observations in a way that reflects the probability that something important *might* have changed. If certain features of the environment have proved in the past to be reliable guides to how changeable it is, we might expect monkeys to be innately tuned to take account of this.

There is indeed a feature of the natural world that varies in colour and is correlated with the risk of change: it's the ambient light from the sky. At dawn and dusk, when monkeys are out foraging for food in a shadowy world where danger lurks and predators are active, *the sky is red.* At midday, when they can relax, *the sky is blue.*

*

I'd moved on in terms of the science. It remained to be seen if the science would help straighten out the philosophy.

I found the results, showing such a neat relationship between colour and frequency of sampling, quite startling. I published them in a well-respected journal and waited for other scientists to take notice. But no one did, and before long I put them to one side. Now, as I return to them many years later, my astonishment hasn't gone away. In fact, having (I hope) matured philosophically, I now find them if anything even more thought-provoking.

This is because, as part of the theory of phenomenal consciousness that I'll be explaining later in the book, I've come round to the view, championed especially by Dennett, that the subjective qualities of sensations must in the final analysis be cashable in terms of *behavioural dispositions.* Not, of course, just one kind of behaviour but as the integral of all the things the sensation motivates the subject to think, and do, and say. The fact that colour experience lines up with such a clear-cut cognitive parameter as sampling frequency bodes well for a future neuroscience of qualia.

By the same token, the results present a challenge that I want to draw attention to straight away. In describing these experiments, I have, so far, skirted round the first-person question: what was it actually *like* to be a monkey sitting in the testing chamber bathed in coloured light?

Monkeys are primates that have eyes and brains much like our own. I'm sure most of us take it for granted that they are sentient creatures whose sensory consciousness resembles ours. It's tempting to assume, therefore, that we humans can guess what it's like for them by imagining what it would be like *for us* if we were in their place. But, not so fast! Two considerations go against this. First, we should not presume we *can* imagine being in the monkeys' place unless and until *we've tried it for ourselves for real.* And so far as I know, no one has. The box was much too small for a person to get inside; and I didn't have the resources to build a man-sized version (of course, I wish I had done). I agree it's possible we'll have tried out other situations that come close. The nearest thing I've come across myself are the coloured rooms created by the American light artist James Turrell. But the very fact that these installations are considered to be art underlines the fact that the sensations we experience in them are unusual and not to be easily guessed on the basis of more everyday encounters.

The second consideration is potentially weightier still. Since the same experiment has not been tried with humans, we can't yet say for a scientific fact whether you or I, if we were tested in the monkeys' place, would feel the urge to get up and move to the other chamber every so often and at a more rapid rate in red light than blue. However, I certainly wouldn't bet on it. And, if we would *not* behave the same way as the monkeys, this would be a strong indication that we would not be having the same subjective experience either. Indeed, if, as Dennett suggests, the quality of the experience is *constituted by* the behavioural dispositions, then our and the monkey's experiences would *necessarily* be different.

Amazing where experiments lead! I say this with feeling because, as things turned out, I was soon to give up doing lab experiments. Towards the end of the 1970s, my monkeys at Madingley started to die one by one from a mysterious illness, called bloat syndrome. An animal I had known for several years would be completely healthy one evening and next morning puffed up with gas and dead. I simply didn't have either the heart or the funds to start over again. In any case, I was being pulled into a different line of research.

9

NATURE'S PSYCHOLOGISTS

Madingley, as I said, was primarily a department of ethology, where people were interested in the behaviour of wild animals. Shortly after I took up my job there, I got to know Dian Fossey, on a visit from her camp in the Virunga mountains of Rwanda where she studied mountain gorillas. Dian was working for a Cambridge PhD, under Robert Hinde, who had earlier supervised Jane Goodall's work with chimpanzees.

Dian was an adventurer more than a scholar. She was out of place and out of sorts in Cambridge. I would find her at the lab, late into the night, gloomily sitting at a desk, chain-smoking cigarettes, and emptying can after can of Coke, while she struggled to turn her field notes into a dissertation that would meet with Hinde's approval. She showed me his red-ink comments on her manuscript. She had written, 'Uncle Bert [a silverback gorilla] charged me, but came to a stop just a few feet away, with an embarrassed look on his face.' Hinde had scrawled acros this, 'How many times have I told you, you really must not use this kind of language.' But, she protested to me, it's the truth, he did have an embarrassed look.

We got talking: I about Helen and monkey brains and consciousness, she about consciousness and gorilla minds and

Uncle Bert. We soon hatched a plan for me to go and visit her for a couple of months when she returned to the field. As it happened, I was between things in my own work and had the time to spare. Mireille had not yet shown me how to tame Helen, and I was waiting for approval of a grant to continue the research on colours. To justify the cost of getting me there, I had to have a project. Dian had an idea. Just before she had left Rwanda to come to Cambridge, a family of the Virunga gorillas had been killed by local villagers. Eight bodies had been recovered and brought back to her camp. Since mountain gorilla skeletons are so rare, it was important that someone should go out and make detailed measurements of the bones, especially the skulls. I was, it's true, not the obvious person to do this. Nonetheless, we pulled some strings and persuaded the Royal Society to provide emergency funding.

I spent a week at the Duckworth Museum in Cambridge, learning the basics of osteometry from anthropologist Colin Groves. I borrowed a set of callipers. And, in early 1971, I was off to catch up with Dian.

*

Her camp, consisting of a few tin huts, was situated at the edge of a grass meadow, in the saddle between two mountains at a height of 3,000 metres. The mountain slopes were covered by forests of ancient Hagenia trees, festooned with trailing vines and orchids. Snow often fell in the night but melted in the morning sun. The night sky to the West was fiery red, lit by a volcanic eruption of Mount Mikeno.

On the climb to the camp, I had already heard gorillas chest-beating in the distance. I could hardly wait to get sight of them. But first, I had my job to do with the bones. The carcases of the dead animals had been stored in plastic sacks. They were rotten and full of worms. I made a cauldron from a kerosene barrel, lit a fire beneath, and boiled the bodies in water for several hours. Then I stripped off the softened flesh, extracted the bones, reassembled the skeletons and laid them out to dry. Over the next days I made measurements of the long bones and skulls.

I hardly expected the measurements to reveal anything unusual. For me, they were simply my ticket to seeing the live animals. But I was in for a surprise. The skulls of four of these individuals turned out to be markedly lopsided, with the left side longer than the right. If mountain gorillas' skulls are asymmetrical, could their brains be asymmetrical too, in the way that humans' are? What implications might this have for gorillas' intelligence? The answers would have to wait until I got back to Cambridge and could consult with Colin Groves. Our paper, 'Asymmetry in mountain gorilla skulls; evidence of lateralized brain function,' was published in *Nature* a year later.[34]

*

Meanwhile, I was hearing things about Dian I would rather not have known. Dian herself had put it about that this group of gorillas had been killed by poachers. Camp gossip, however, told another story. Talking to the men who worked for her, I heard the villagers' version. Dian, believing that poachers hunting bushmeat had come from their village, had decided to wage war. She had dressed up in a Halloween mask and raided the village, capturing a young boy, whom she held for several days. The villagers, outraged and humiliated, took the only revenge available to them: they killed the creatures she seemed to care about above all.

I soon began to see the dark side of Dian for myself. Her bravery was legendary but so was her cruelty. She was an unapologetic racist, who spoke in derogatory terms of the Rwandan men she

Figure 9.1 The author and Dian Fossey. Gisenyi, Rwanda, 1971.

hired to assist her and showed them no respect. She could be malevolent to whites as well: students, colleagues, anyone who crossed her. 'Guess what?', she wrote to me in a letter when I was back in England, 'A European tourist got chewed up into a hamburger by a silverback of Group 6. Needless to say, I have little sympathy.'

She drank heavily. In camp, I saw, to my dismay, how the soulful, considerate, loveable Dian I had met in Cambridge could become a ranting whisky-fuelled Lady Macbeth on her home ground. For everyone around her, it was a constant dance to anticipate her moods and stay on the right side.[35]

*

Still, thankfully, these problems weren't so obvious to begin with. Two weeks in, I was done with the bones and ready to go out and observe the living animals.

The plan was to set off at dawn to track down a particular group and spend the day, even sometimes the night, with them. On the first outings, I went with an experienced Rwandan guide. But I soon realized that having a guide was constraining and not necessary. Gorillas leave an obvious enough trail as they move through the forest, and they aren't hard to follow. Without a human companion, I could enter the gorillas' space on my own terms.

I would construct a day-nest for myself on the periphery of the group, settle in, grab myself a handful of vegetation to chew on, and wait to see what would come to pass.

In that fairy-tale place, watching and watched by a dozen black eyes, wondering about their next move and they no doubt wondering about mine, I became caught up with questions about *inter-subjectivity* and *social understanding*.

*

It's hard to believe, nearly fifty years later, that the idea of *social intelligence* could have been at that time a novel one. But in my academic education up to that point, I may say I had never heard the words 'social' and 'intelligent' mentioned in the same breath.

Intelligence, so I'd been taught, had to do with the capacity to find answers to clear-cut physical or mathematical questions, not messy social ones. Now, however, two considerations were forcing these concepts together. First, these gorillas—as had been made clear to me from measuring their skulls—had huge brains, much larger than any of the other forest animals. Large brains surely mean high intelligence and skill at problem solving. Second, however, as I could see from observing their behaviour, their life in the forest presented them with very few problems that actually needed solving. Food was abundant and easy to harvest and there was no danger from predators—little to do, in fact (and little done) but eat, sleep, and play.

From an evolutionary standpoint, this didn't make sense. Why go to the trouble of developing a large—and energetically expensive—brain if it's not needed? I realized the answer had to be that there was more to the gorillas' apparently uncomplicated lifestyle than I was seeing. Was there something going on, perhaps right before my eyes, that I was simply failing to pick up—or at least not recognizing as requiring high intelligence?

I tried to put myself in the gorillas' place and to imagine what—if anything—might really tax their minds. But I found myself thinking equally about myself. Where did my real problems lie? At home in Cambridge, in the camp with Dian…the fact was, I never let up thinking about *my relationships with other people.* If I do this, what will she do? But suppose I did that or if she did something else?

Then it dawned on me: for these gorillas too, their problems must primarily be social ones.

*

The reason why day-to-day subsistence poses so few problems for the gorillas is precisely because the gorilla family as a social unit is so well adapted to it. An infant brought up under the care of others, protected while it's young, shown the ways of the forest, will have no difficulty with the practical problems of survival. However, the need to maintain relationships within the family group may present problems of quite a different order.

The family life of gorillas may not look, to an outside observer, all that problematical. But that's only because the animals themselves are so accomplished at it. They know each other intimately. They know their place. There are, of course, frequent small disputes: about who grooms who, who should have first access to a favourite food, or sleep in the best site. But for the most part these are quickly resolved. Sometimes, however, the disputes will be more serious: major disagreements about social dominance, the right to mate, whether a young male should be turned out of the family, whether a strange female should be allowed to join them. These power struggles may not happen often, but when they do they can be literally a matter of life and death. Most silverback gorillas bear scars, received in brutal fights. Most older females have lost at least one baby, killed by a male.

The challenge for every individual gorilla will be to do well for themselves while preserving the social network on which they all depend. The winners will be those who are best able to read the signs and anticipate how others are going to behave—so as to help them, outwit them, or manipulate them. The better they are at understanding the minds of other gorillas, the more likely they will be to pass on their own genes.

In short, gorillas need to be by nature skilled psychologists: 'natural psychologists'. *That*, I realized, is what will have driven the evolution of intelligence. Doing psychology could require every ounce of a gorilla's brain power that's available.

*

What does this say about human beings? There's no question that for us, too, the most promising but also the most dangerous elements in our environment are other humans. Once our ancestors left the forests and began to live as cooperative hunter gatherers on the savannah, they depended more than ever on their ability to read the minds of those they lived alongside.

Human social arrangements have evolved to be orders of magnitude more complex than those of our ancestors among the apes. And, indeed, the human brain has gone on growing. It's now more than three times the size of a gorilla's. Human intelligence is commensurately greater too.

The story fits. I wrote up these ideas in a paper entitled 'The Social Function of Intellect'.[36] Other researchers took up the ideas and ran with them. Fifteen years later, Robin Dunbar recast my 'social intelligence hypothesis' as the 'social brain hypothesis'. In Dunbar's view, brain size alone determines how many relationships an individual can keep track of—so that while gorillas max out at about 15, humans can manage up to 150 friends ('Dunbar's number').[37]

I had suggested the link to brain size myself. However, I may say I soon became less inclined to put as much emphasis as Dunbar has done on size as such. It began to seem too simplistic. I could hardly believe that what's required for doing psychology is the raw computational capacity provided by extra brain cells. Whether you are human or gorilla, surely, it's going to be the *way you think* about other individuals—what *kind of actors* you imagine them to be—that will be crucial to your ability to predict their behaviour and manage your relationships.

So, as I sat in my day-nest, wondering what it was like to be a gorilla, wondering what the gorilla might be wondering about what it was like to be me, I began to home in on the essential role of *introspective consciousness*. Over and above *brain power*, a natural psychologist needs a *brain story*. And this, it would seem, is just what consciousness helps to provide: a story couched not in terms of brain states but as a user-friendly narrative about conscious experiences.

Look at us humans. Mind-reading, as humans practice it, revolves around self-knowledge. We discover by introspection the intimate story of our selves. Then, when we want to model the mind of someone else, we construct the other's mind in the image of our own. We assume the other to be a conscious subject who thinks and feels in the way we've learned that we do. Then we read into them the mental states that *we* would have if we were in their place, and we expect the thoughts and actions that flow from these states to follow the same path as they would for us. We can do this because—but only because—we've experienced these very states of mind ourselves and seen for ourselves how they connect.

It can be very simple. You see someone prick their finger and feel their pain yourself. You see them reach for an umbrella and assume they think it's going to rain . . . It can be very complicated. You want to punish Dian for kidnapping your son; so you plan to kill one of her gorillas, to hurt her as she hurt you.

Do gorillas think like this? Not about kidnaps but about pain and rain? Very probably. And about punishment? Maybe that too.

*

But back to the essentials. The gorillas and I were *watching* each other. It may seem too obvious to mention, but I assumed that behind those eyes were conscious creatures who were *seeing*. And that seeing for them was very much like what it was for me. I was thinking about how things *look* to them.

Then I made a long-distance connection. At home in Cambridge, a few months previously, I'd been trying to imagine what it was like to be my monkey, Helen. With her, I'd found that, if I assumed her visual experience was similar to mine, I couldn't get far. I was simply unable to explain anomalies in her behaviour, for example, how she'd behave when she was stressed. However, if Helen had blindsight, then in truth her experience wasn't at all like mine. No wonder my attempt at mind-reading failed.

So, here's a thought. Suppose Helen were with me watching the gorillas, and *she* were trying to imagine what it was like to be them. Would she, with her sensationless vision, realize that behind those eyes were conscious creatures who were *seeing* the world around them? Probably not.

Things were coming together. A couple of years later, I would write a paper entitled 'Nature's Psychologists', where I discussed how the ability to mind-read depends on shared conscious experience. With regard to this scenario with Helen:

> I believe that Helen's lack of visual consciousness would have shown up in the way she herself conceived of the visually guided behaviour of other animals—in the way she did psychology... Being blind to the sensations of sight, she would be blind to the idea that another monkey can see.[38]

10

ON THE TRACK OF SENSATIONS

To work.

I've described in the previous pages what may appear to be the rather random walk of my first years as a researcher. There was, however, a common theme. From one direction or another, I kept coming back to questions about consciousness—in particular, the philosophically provocative nature of sensations. Phosphenes, blindsight, aesthetic preferences, social intelligence, theory of mind, even religious awe...each of these encounters gave me a different perspective on the problem and left me, as I see it today, with a piece or two of the solution.

I now believe that a rather simple evolutionary theory can tie it all together. But before coming to this in Chapter 11, I want to prepare the ground by discussing some of the philosophical issues that the theory must either sidestep or resolve. You may consider some of this to be unduly fussy. I might even agree with you. I hope you'll see why we need to go there all the same.

*

Sensations as representations

We've wandered some way from the definitions of sensation and perception I gave earlier in the book. But now we must return to basics. Sensations and perceptions are psychological events. They are *ideas* you form, on the basis of sensory information, about what's happening around you. Sensations are about what's happening to you at your sense organs. Perceptions are about the state of the world.

In the language of cognitive science, sensation and perception both involve 'representing' by the brain. Representing is an active process that involves making a 'representation' that is *about* something *for* someone. The thing that constitutes the representation is called the 'vehicle'; the thing the representation is about is called the 'representandum'; the someone the representation is for is called the 'representee'. Thus, for instance, the spoken word DOG can be a representation where the sound of the word (the vehicle) represents the idea of the particular animal (the representandum) to a speaker of English (the representee). Note that there's no such thing as a free-floating representation. Taken out of context, the sound *DOG* doesn't represent anything at all.

Representations are designed to match the representee's interests. This means that, if there are representees with different interests, the same facts may be covered by representations with different representandums. For example, the London Underground railway can be represented in several ways. One way, the familiar Tube Map, is designed for travellers. It shows the names of the stations, their relative positions on a schematic diagram, the connecting lines, the stations where it's possible to transfer, the fare zones, etc. But another map is designed for engineers. This shows the exact geography of the tunnels, the depth below ground at every point, the location of ventilation shafts, drainage sumps, electrical supply lines, etc.

It's rather the same with perception and sensation. They both begin with the data about stimuli arriving at the sense organs—light at your eyes, sound at your ears, pressure on the skin, and so on. But your brain then forms those two separate representations to serve your interests on two different levels. On the one hand, it creates a sensory representation that allows you to track the nature of the stimulation at your body surface and how it's affecting you: the sweet taste is on *your* tongue and enticing, the shrill sound is at *your* ears and upsetting. On the other hand, it creates a perceptual representation that allows you to track the features of objects in the external world: the sweet taste is the taste of honey, the sound is the sound of a crying baby.

The key thing about sensation, in contrast to perception, is that it is essentially body-centred, evaluative, and personal. It's almost as if you are expressing your bodily opinion about what the stimulation means to you: responding with a kind of inner smile, or wince, or frown. Indeed, as we'll see shortly, there's a solid reason it feels this way. I'll be arguing, as a central plank of the theory, that the vehicle for the brain's representation of sensation is in fact a form of covert bodily expression. You respond to sensory stimulation with the *makings* of an action—never completed—appropriate to what's happening and how you feel about it. And then you *read your own response* so as to get a mental picture of it.

Phenomenal properties: real or illusory?

Whatever the brain actually does to represent sensation, the vehicle is presumably some kind of nerve cell activity. There won't, at this stage, be anything that can't be described in physical terms. It's when you form the idea of the representandum that things begin to get tricky. You describe your sensory experience to yourself in terms of phenomenal qualities—redness, painfulness, sweetness, and so on—that seem to have no counterpart in physical reality. Your sensations are nowhere in physical space; they even seem to occupy a time of their own, 'thick time' that outlasts the physical instant.

For theorists, of course, this sets alarm bells ringing. If a physical process could not have these phenomenal properties, then how can sensations have them? Does it mean they are some kind of illusion, a trick of the imagination? This has become a contentious issue in the philosophy of consciousness. Proponents of 'illusionism', Dan Dennett and Keith Frankish chief among them, maintain that phenomenal properties are indeed pure make-believe. There's nothing in the real world that does or could actually instantiate these properties.[39]

This was my position too until quite recently. I liked to draw an explicit parallel between having a sensation of red, for example, and looking at the 'real impossible triangle', the wooden object devised by Richard Gregory, shown in Figure 10.1.[40] When you look at this object from the critical position, as on the left, you

Figure 10.1 The Gregundrum

represent it as having impossible physical properties. Likewise, I suggested, when you experience red light reaching your eye, you represent it as having properties it couldn't really have.

I'm no longer convinced by this analogy. The crucial question with sensations is: what exactly is the representandum? What are you attaching these unreal properties *to*?

As we saw above, sensations track the nature of the stimulation at your body surface and how it's affecting you. So, the red sensation does indeed represent the light at your eye, but it does more than this; it represents *how you feel* about the light at your eye. And that's what makes the analogy with the Gregundrum less than helpful. For sensory feelings are nothing like wooden triangles, even weird ones. They are not constrained by the laws of physics. They are your *idea* of what it feels like to have this happening to you. And, as such, they can have whatever properties have proved in the course of evolution to be appropriate to describing this subjective state. If these properties turn out to be non-physical or even para-physical, that's just what we might expect. It doesn't mean that these phenomenal properties should be written off as invalid or 'illusory'. Instead, we should welcome them for *what they are* and *what they do for your sense of your own being* (which, of course, our theory is going to have to explain).[41]

Phenomenal properties: out there or in here?

Oscar Wilde wrote, 'It is in the brain that the poppy is red, that the apple is odorous, that the skylark sings.'[42] Most people know this, of course. Over a lifetime of playing around with their own sense organs (pressing on their eyeballs, maybe!), they have come across plenty of evidence that phenomenal properties are their own subjective creation.

The fact remains, however, that, even while you may understand quite well that sensations are essentially about how *you relate* to external stimuli, you may harbour another kind of illusion. You may find yourself casually believing that the properties you ascribe to your personal sensations actually inhere in the external objects you perceive: that the poppy itself is phenomenally red, the apple phenomenally odorous, the skylark's voice phenomenally sonorous.

The philosopher David Hume made much of this:

> Tis a common observation, that the mind has a great propensity to spread itself on external objects, and to conjoin with them any internal impressions...Thus as certain sounds and smells are always found to attend certain visible objects, we naturally imagine a conjunction, even in place, betwixt the objects and qualities, though the qualities be of such a nature as to admit of no such conjunction, and really exist no where.[43]

Hume calls this out as a logical mistake. Likewise Reid:

> The perception and its corresponding sensation are produced at the same time. In our experience we never find them disjoined. Hence, we are led to consider them as one thing, to give them one name, and to confound their different attributes. It becomes very difficult to separate them in thought, to attend to each by itself, and to attribute nothing to it which belongs to the other.[44]

Theorists in general have gone along with Hume and Reid in regarding this kind of projection as a conceptual error, a misattribution that may not matter much but has nothing to recommend it. I, as an evolutionist, however, tend to assume that if and when humans have 'a great propensity' to get things wrong, there's likely to be a plus side to it. And, indeed, I think the philosophical sages have generally underestimated the active interest we humans have in making this particular misattribution.

Here's the important consideration. If your encounter with a particular object out there in the world gives rise to a particular sensation in you, it's very likely it will arouse a similar sensation in anyone else who interacts with it (and, of course, yourself another time). If the poppy arouses a red sensation when you look at it, it will do the same for another person. Therefore, the poppy, we might say, is 'rubro-potent'—potentially red-sensation arousing for humans. By the same token, a sugar-lump is dulci-potent. An ice-cube is frigi-potent. A rose is fragra-potent. So, while you'd still be wrong to say the poppy *is* phenomenally red, you'd be right enough to say the poppy *looks* phenomenally red, the sugar *tastes* phenomenally sweet, the ice cube *feels* phenomenally cold, the rose *smells* phenomenally fragrant.

These potencies are genuine—even if secondary—properties of the objects to which you assign them. Every time you project your sensation onto an object of perception, you are anticipating, correctly, what it will be like for another person to encounter it. The poppy you *perceive* to be physically red becomes at the same time a poppy you *feel* to be phenomenally red for humans. Cold perception is getting overlayed—hijacked even—by hot, subjectively shared sensation.

*

There's a story I've been wanting to share. A painter of my acquaintance, Sargy Mann, having suffered years of deteriorating sight, became all of a sudden totally blind. A few days later, he was mooching round his studio, wondering what he would do with the rest of his life, and he realized the one thing he still wanted to do was paint. In his words:

> After a bit I thought: 'Well here goes', and loaded a brush with ultramarine. What followed was one of the strangest sensations of my life: I 'saw' the canvas turn blue as I put the paint down. Next I put down my Schminke magenta, and 'saw' it

> turn rose. The colour sensation didn't last, it was only there while I was putting the paint down, but it went on happening with different colours.[45]

My take—for what it's worth; you may have a better one—is this. It's the job of a painter to show people new ways of seeing the world. To do this, he has to *gain control* of their visual consciousness. He puts down paint in order to bring their sensations into line with his own artistic conception. In the words of Robert Browning, 'Art was given for that; God uses us to help each other so, Lending our minds out.'[46] So, in those poignant first moments of sightless painting, Mann was recreating the colour sensations he still hoped to lend to others.

The hard problem: the explanatory gap

The next problem may be the hard one. Assuming that sensations, with their phenomenal properties, are feelings generated by the brain, as philosopher Dan Lloyd has put it, we want 'a transparent theory' of how this works: 'one that, once you get it, you see that anything built like this will have this particular conscious experience'.[47]

The question is, how we can get to such a theory, even in principle. How can anything built of unconscious physical bricks conceivably give rise to a conscious edifice? The objection that, in one form or another, gets raised again and again is that physical matter is *inadequate as a cause* to produce phenomenal consciousness as an *effect*. Nothing will come of nothing—and, where consciousness is concerned, the physical brain is nothing.

The argument goes back to René Descartes, although he didn't immediately apply it to the case of consciousness. In the third of his *Meditations,* he questioned where his idea of God had come from. He realized he was able to conceive of God as a *perfect being.* Yet, he reasoned, the idea of perfection could not be assembled from less-than-perfect components; and nothing about his own thought could be perfect. Therefore, he concluded with a flourish, the idea must have been implanted in him directly from above:

> I recognize that it would be impossible for me to exist—having within me the idea of God—were it not the case that God really existed. By 'God' I mean the very being the idea of whom is within me, that is, the possessor of all the perfections which I cannot grasp, but can somehow reach in my thought.[48]

This argument as it relates to God—that it takes perfection to create perfection—may not strike you (as it has not struck most later philosophers) as all that convincing. But when the parallel argument is made, that it takes consciousness to create consciousness, it has seemed harder to deny its considerable force. Phenomenal consciousness has those strange other-worldly properties. You cannot make something other-worldly out of worldly materials. But worldly materials are all your physical brain has to work with. So, consciousness cannot have been constructed by the physical brain.

As philosopher Colin McGinn has colourfully put it:

> You might as well assert, without further explanation, that space emerges from time, or numbers from biscuits, or ethics from rhubarb…Matter is just the wrong kind of thing to give birth to consciousness. How can physical properties of the brain generate phenomenal features?…The physical brain just doesn't have the resources to do the kind of generative work you are asking of it: it's not a miracle-box.[49]

Alfred Russel Wallace, co-discoverer with Darwin of evolution by natural selection, used this argument, rather as Descartes did, as an excuse for slipping in an Intelligent Designer by the back door:

> No physiologist or philosopher has yet ventured to propound an intelligible theory, of how sensation may possibly be a product of [material] organization; while many have declared the passage from matter to mind to be inconceivable... You cannot have, in the whole, what does not exist in any of the parts... The inference I would draw from this class of phenomena is, that a superior intelligence has guided the development of man in a definite direction.[50]

Other theorists, the Panpsychists, draw a still more radical inference. If it is inconceivable that consciousness could emerge from non-conscious matter, and yet it does emerge from the material of the brain, this has to mean the matter of the brain is conscious to begin with. In fact, possibly all the matter in the universe is conscious to a small degree. Philip Goff, an enthusiastic proponent of panpsychism, relies on an argument that mirrors that of Descartes: consciousness, he says, is defined by its *qualities* and you can't get quality from quantity. 'To say that reality can be described in purely *quantitative* terms is to say there are no *qualitative* properties.'[51]

I'll have more to say about panpsychism in Chapter 16. For now, I'll simply state my opinion, that it can't deliver what it claims to. No version of it even comes close to providing a 'transparent theory' of the kind Lloyd called for: a theory that would allow us to see that anything built like this will have this particular conscious experience. Rather, panpsychism gratuitously postulates that consciousness is present wherever it is in fact present. And, in doing so, it leaves us none the wiser. As Bertrand Russell once remarked, 'The method of "postulating" what we want has many advantages; they are the same as the advantages of theft over honest toil.'[52]

Nonetheless, I think we should recognise that the principle of sufficient causation deserves respect. Causes must indeed be adequate to their effects. You can't pull yourself up by your own bootstraps; you can't have more information in the output of a communication channel than is present in the input; you can't necessarily deduce all the truths of mathematics from its axioms. And the examples could go on.

So, there could be a problem here—'an explanatory gap' between the brain and the phenomenal properties of sensations. There *could be*. But only if we cleave to the idea that sensations must somehow be *identical* to brain states. And, with regard to this, the waters have been muddied by what has seemed to many researchers to be a good proposal. It's that if we are to arrive at a transparent theory, we must discover the *neural correlates of consciousness*, the NCC.

The NCC was defined by Francis Crick and Christof Koch as 'the minimal neural mechanisms that are necessary and sufficient for the occurrence of a particular conscious experience'. Fair enough, you might think: just the kind of concept we need to get the science of consciousness headed in the right direction. Yet, the trouble is that people have taken it to mean that what we're looking for is some process in the brain that actually *has* the properties of the conscious experience. 'Necessary and sufficient' sounds like the full deal. So, the presumption is that when you see red, for example, there is a neural correlate that possesses the very property of phenomenal redness.

But this presumption, as must now be obvious, is all wrong. When you see red, there won't be any activity *of* the brain that *is* phenomenally red; there will only be some activity *by* the brain that *creates the idea* of phenomenal redness. Hence, what we should be looking for is not the neural correlates of consciousness but the neural correlates of *representing* consciousness, not the NCC but the NCRC.

This means, in turn, that we should be looking for a two-stage process. First, there will be the brain activity that is the *vehicle* for the representation. Then, quite separately, there will the brain activity that takes this vehicle to *point* to the idea. And there's no reason whatever to expect either of these brain processes to have phenomenal properties in its own right.

A new analogy may help here. Rather than thinking about the Tube Map, let's consider the text of the novel *Moby Dick*. Suppose we want to explain how someone who encounters the printed text gets to conjure up a mental picture of the great white whale. First, we must explain how the *text* tells the story. Then, how a reader of English *makes sense* of it. But, of course, neither the text nor the sense-making has to be white or whale-like.

Then the problem of insufficient causation no longer looms so large. True, Descartes believed that ideas, as much as anything else, require adequate causes. Maybe Goff believes something like this too: that you can't even get the *idea* of quality from quantity. But there's really no reason to take this seriously. The fact is, both common sense and example contradict it. A finite brain can evidently generate the idea of infinity. An amoral brain can generate ideas of truth, beauty, and goodness. A worthless brain can generate ideas of value. Or, to jump to a level that perhaps brings us closer to phenomenal properties, a brain that obeys classical physical laws can come up with quantum theory. McGinn says the brain is not a miracle box: but actually, in the realm of ideas, that's just what it is.[53]

If there is, after all, no unbridgeable explanatory gap, the hard problem of consciousness becomes just an ordinary scientific problem.

The hard question: And then what happens?

Time to move on to what Dennett likes to call the 'hard question'. It's '*And then what happens?*' 'The question, more specifically, is: once some item or content "enters consciousness", what does this cause or enable or modify?'[54]

I'd remark that this isn't a question we can or should *always* ask about the workings of a natural phenomenon. Once we have an explanation, for example, of how a rainbow works, how sunlight is diffracted by raindrops, to produce the coloured arcs, we're not obliged to ask what happens next (unless, of course, we want to find the legendary crock of gold). With consciousness, however, there are two considerations that make it the right question. First, phenomenal consciousness is, in the first instance anyway, a *private* state of mind. This means that we only know about it in others through its downstream consequences on the way they act—what they think and say and do. Second, we have every reason to believe that phenomenal consciousness has *evolved by natural selection*. This means that whatever the downstream consequences are, taken as a whole they must be having some kind of positive effect on how individuals live their lives—and ultimately on their chances of biological survival (the Darwinian crock of gold.)

Of course, before we can ask what happens next, we need to know what has happened already so as to bring sensations with phenomenal properties to consciousness. Let's presume it works the way we've suggested above. Your brain creates a representation of stimuli arriving at your sense organs, and you, the representee, read this in order to arrive at the *idea of what the stimulation feels like*. So, the next question is: what does this idea—the sensation with its phenomenal properties—'cause or enable or modify'?

The short answer is that, if being conscious of feeling this way about the situation is to show up in behaviour, it must cause changes in your mental attitudes that dispose you to act in ways you wouldn't have done otherwise—the mental attitudes being the beliefs, hopes, and so on that you entertain about the situation or about yourself.

Think back to the apparition on the church in Knock. Archdeacon Cavanagh arranged that a lantern slide was projected onto the gable wall. The villagers took this to be an apparition of Mary and her saints. And then what happened? They took it to be a God-given miracle, and this led them to prostrate themselves in gratitude. As Bridget Trench testified, 'When I arrived there I saw distinctly the three figures, I threw myself on my knees and exclaimed: "A hundred thousands thanks to God and to the glorious Virgin that has given us this manifestation."' As Dominick Byrne said, 'The night was dark and raining, and yet these images, in the dark night, appeared with bright lights as plain as under the noonday sun. I was filled with wonder at the sight I saw. I was so affected that I shed tears. I continued looking on for fully an hour.'[55]

So, if the attitudes that flow from seeing a saintly apparition dispose you to throw yourself to your knees and cry, what might the attitudes that flow from phenomenal consciousness dispose you to do?

Sensations can modify action-relevant attitudes on many levels, directly and immediately or through a complex chain of other ideas. When you represent what it's like as painful or pleasurable, the low-level recognition that this is bad or good for you may well be the most important factor determining what happens next. You scratch the itch; you relax in the warmth of a bath. But sensations with phenomenal properties can lead to life-changing beliefs on an altogether different level—in humans, all the way through to thoughts about the 'soul'.

The question is, at what level do the *phenomenal properties as such* become relevant?

We should think twice about this. We shouldn't assume it's at the lowest level. The phenomenal quality of a sensation is actually neither necessary nor relevant to basic judgments about what's good or bad for you. In the case of pain, for example, when you touch the stove you could very well represent the sensory event as bad and withdraw your hand even if your sensation didn't have any phenomenal dimension to it. As we discussed in Chapter 1, the phenomenality that as a human you take for granted is a feature of sensory representations that has been added in the course of evolution. For us, it's hard to imagine generic pain without the phenomenality and maybe especially hard to imagine non-phenomenal pain being *bad*. Yet this must be how it was for our insentient ancestors and still is with creatures that have remained insentient to this day.

Let's consider what it's like to be a frog. If a stinging chemical is applied to a frog's skin, it will take action to scrape it off. And, indeed, it will do so even after its brain has been destroyed. T. H. Huxley, lecturing in 1870 on 'Has a Frog a Soul?' described the experiment:

> Suppose that the head of a frog be cut off so as to detach the whole brain... If the frog be laid on its back, it will remain passively in that position. If one of the feet be touched with acid, the leg will be retracted, and then the two legs will be rubbed together to get rid of the irritating matter. Not only so, but if the irritated limb is placed in an unusual position, for example, drawn up at a right angle to the body, the other leg will be gradually raised up into a corresponding position, until it is so

> placed that it can rub away the irritating matter. Here is evidence that...it has a power of adjustment which enables it to meet an entirely new case—to solve a problem which could not have been presented to it under the ordinary conditions of the life of the frog.[56]

Leaving aside the question of what it's like to be a normal frog, let's presume that a frog without a brain doesn't experience pain on a phenomenal level. Yet, it undoubtedly shows typical pain behaviour.

And it's not just frogs. Mammals, including humans, still show many signs that we normally associate with phenomenal experience even when the cerebral cortex is missing. Hydrancephaly is a rare condition of human children in which the brain's cerebral hemispheres are absent and replaced by sacs filled with cerebrospinal fluid. Nonetheless, as reported by Bjorn Merker:

> hydrancepahalic children show emotional or orienting reactions to environmental events...They express pleasure by smiling and laughter, and aversion by 'fussing', arching of the back and crying (in many gradations), their faces being animated by these emotional states...They show preferences for certain situations and stimuli over others...And their behaviours are accompanied by situationally appropriate signs of pleasure or excitement.[57]

Psychologist Mark Solms, reviewing this evidence, states: 'One surely must conclude that it does feel like something to be these children.'[58] In other words, they are phenomenally conscious. But I suggest we should draw quite the opposite conclusion. Yes, for sure these children are representing what's happening to them as good or bad. But in the absence of higher brain centres these representations are generic sensations without phenomenal properties. It is *not* 'like something' to be these children—or these frogs. And if this is how it is with hydrancephalic children, why not with human beings with normal brains? Typical pain behaviour, such as withdrawing your hand from the stove, probably has little if anything to do with the phenomenal quality of the experience.

As we'll see when we come to a fuller discussion of the issue later, the level at which phenomenal quality does become relevant is not so much in your beliefs about the stimuli affecting you as in your beliefs about your *self*: that *you* are the being having the experience. And at this level, the phenomenality of sensations, as such, can become the central fact. 'Gosh, look what I have here!' It can even lead humans to reclassify pain from bad to good. The climber Joe Simpson, in his remarkable story of surviving a terrible fall in the Andes, describes how, after briefly losing consciousness, when he came to: 'A burning, searing agony reached up from my leg. It was bent beneath me. As the burning increased so the sense of living became fact. Heck! I couldn't be dead and feel that! It kept burning, and I laughed—Alive! Well fuck me!—and laughed again, a real happy laugh.'[59]

Dennett's question—'And then what happens?'—opens windows onto areas where most consciousness theorists have hardly thought to look. It's a question to which, as things stand, we don't have anything like a complete answer for pain or any other sensation. The gamut of effects that being phenomenally conscious has on you must include everything you think and say and do as a result: the crucial effects it has on your sense of self; the changes it produces in emotion and mood; all the memories it brings out; all the influences on your enjoyment of life and self esteem; and, of course, all the provocative thoughts you may have about how mysterious, and inexplicable it is—including the embrace of dualism, panpsychism, and related philosophical extravagances.

Suppose we could actually discover *everything* that happens next at the level of brain and behaviour, would we then have discovered *everything there is to know* about the content of phenomenal experience? To revert to Descartes's way of putting things, would we find the consequences have 'adequate reality' to fix the experience? Dennett is firmly of the opinion that they would. I'm inclined to agree with him. But this makes him and me relative rarities. A great deal of ink has been spilt by philosophers in arguing that the first-person experience of sensations can never be captured in its entirety by any such public third-person description.

'What Mary didn't know': the knowledge argument

In a famous thought-experiment, Frank Jackson dreamed up the case of Mary.[60] Mary is a brilliant scientist who studies how colour is represented in the human brain and its psychological effects but who has been forced to investigate the topic from a black-and-white room via a black-and-white television monitor. Although she herself has never seen coloured light, she has discovered everything that goes on in the brain when, for example, a person looks at a red wall and has a red sensation; moreover, she has observed all the sequelae at the level of thought and behaviour. What's more, she has written out an objective description of all this that can be shared with other scientists.

Now, the time comes when Mary is released from the room. She emerges into a world of colour and sees red for herself for the first time. The question is: does she learn something essential about seeing colours she didn't know before? Does she now *know* something new?

You may think it's obvious. Yes. Mary now knows, from the first-person standpoint, 'what it's like to see red'—and this knowledge simply could not be derived from her painstaking research as an outsider. That's to say, if Mary had previously been so scientifically arrogant as to imagine she had arrived at a transparent theory of consciousness, she'd now have her comeuppance. She never realized that a brain built as hers is would have the particular conscious experience she's now having.

The 'knowledge argument', as it's often called, is regularly touted as a knock-down philosophical argument against the possibility of a materialist theory of consciousness. Goff, for example, makes it the cornerstone of his book promoting panpsychism.

Yet, I want to say: not so fast. What exactly are you claiming when you say that Mary gets to *know what it's like*? I expect your recourse will be to take yourself as an example. *You* know what it's like for *you*, and Mary now knows what it's like for Mary in just the same way. That's fair enough. But you're begging the question. What is it that you—and Mary—know about yourselves? I suspect the truth is *you know very much less than you suppose.*

When you look at a poppy, you may well say you know that having a red sensation is *like this*. You can point to it as a mental state of your own. But that's a weak kind of knowledge that relies on pointing to an ongoing experience. While it's there for you, the experience appears to have a rich phenomenal structure. But how much of this will you be able to summon up offline? The fact is that as soon as the sensation is no longer before you (when you close your eyes, say), your knowledge of what it was like immediately shrinks. In fact, all you are left with is a kind of afterglow of what it made you feel like thinking and doing.

In that case, why should we suppose that there was ever more to it than this? I think we should be highly sceptical of the suggestion that, while you are in process of having a red sensation, you know so much more about what it's like than you can show for it later. Surely, the safer and more economical assumption must be that, even at the time, your knowing was constituted and exhausted by the attitudes you carry forward: those very attitudes that have the public consequences we've just discussed.

I agree it doesn't seem like this. The impression you get is that there is a side to phenomenal experience that, when you try to pin it down, continually gives you the slip. One of the peculiar things about sensations is that the present moment, the 'now' of sensation, has a paradoxical dimension of temporal depth. Each instance of sensation seems to hang on a little, as if it happens for longer than it happens. The result is that sensations seem to be on the edge of having a more permanent existence which is belied by the reality. Robbie Burns caught the truth of it in verse. Moments of presence are 'like poppies spread, You seize the flower, it's bloom is shed; Or, like the snow-fall in the river, A moment white, then melts forever.'[61]

We resist recognizing this for a good psychological reason. We'll see later how the experience of sensations is essential to scaffolding a sense of self. But, if this scaffold is always on the brink of vanishing, it requires constant reaffirmation. The poet Coleridge describes how his three-year-old son awoke during the night and called out to his mother. 'Touch me, only touch me with your finger.' 'Why?', she asked. 'I'm not here', the boy cried. 'Touch me, Mother, so that I may be here.'[62] No wonder we want to imagine there's more to being here than there is.[63]

But back to the Mary experiment. I think these considerations turn the argument around. If 'knowing what it's like' to have a particular phenomenal experience amounts to no more or less than knowing the full range of attitudes the experience gives rise to, then Mary will surely have discovered everything about this from her scientific studies of what happens next for other humans. In fact, we should expect Mary to be ahead of the game.

As an expert psychologist, she will have seen it all at the level of behaviour; as a neuroscientist, she'll have seen it at the level of brain activity. So, she'll know everything you yourself know and more.

It follows that when Mary first sees red she will be epistemically totally prepared. She will not gain any *new knowledge* from the new experience. This isn't to say it won't be a new experience for her. Of course it will. As you can no doubt attest in your own case, you don't actually get to see red by knowing what it's like. Mary has never got to see red before. Nonetheless, the attitudes about seeing red she finds herself having as a consequence of her experience will be exactly those she would have predicted.

Do you still have a nagging intuition that this can't be right? While Mary is actually having the first experience, won't there be *something* about it she won't have anticipated—albeit something that is totally ephemeral and ineffectual? Something she knows about when she's inwardly pointing to it, but leaves nothing behind? . . . The intuition is widely shared. It may indeed contribute to your sense of just how special it is to be *you*—the momentary possessor of this inexorably private, vanishing experience. But as an argument it vanishes too.

'What natural selection didn't know.' The inverted spectrum

You might think that the Mary story is merely a thought-experiment that has no bearing on real life. But in fact it clearly has a major bearing on the *evolution* of real life. For if we were to be persuaded by the discussion of Mary that there are essential features of phenomenal consciousness that could not in principle be discovered through outside scientific observation, we would be hard put to explain how these features could have evolved. If a scientist like Mary couldn't know about a particular feature, nor could natural selection. It's reassuring, therefore, that this potential problem for an evolutionary theory of sentience can be blocked.

Let's note, however, that we mustn't presume it works the other way round: that if a scientist like Mary could see a particular feature, natural selection must be able to see it too. Mary, remember, is able to see everything about both brain and behaviour, while natural selection can 'see' only behaviour that impacts biological survival. So, now the interesting question is: what about *you*? When you have a red sensation, is your first-hand knowledge of what it's like closer to what Mary knows or to what natural selection knows?

You certainly know less than Mary about how the sensation is being represented in your brain. But we can presume you know just as much as she does about what happens next, that is, the beliefs and attitudes that result and their behavioural consequences. Moreover, this will be knowledge of *all* the consequences. Natural selection's knowledge, however, must be limited to those that affect survival. Thus, it seems very likely that, while you know less than Mary, you actually know more than natural selection has ever done. This is potentially significant because it opens up the possibility that phenomenal experience could have evolved in idiosyncratic ways unseen by natural selection, what's more that there could be inherited individual differences in what it's like.

Let's suppose that the brain can in fact represent the phenomenal quality of a sensation of red light in significantly different ways. These differences in the brain will make a difference to subjective experience and so a difference to what happens next at the level of behaviour. However, it's quite possible that, in the areas in which natural selection takes an interest, these differences, though real enough to the subject, are irrelevant. All versions get the job done. Then there would be no reason for natural selection to have favoured one variant of the experience over another. So, it will have been a toss-up which of several have become established.

Think again of the example of the apparition on the church in Knock. Archdeacon Cavanagh could have used any of many different lantern slides depicting the Virgin and her saints to create the apparition. The villagers would have had different visual experiences with different slides. Yet, each of the slides could have evoked equivalent feelings of religious awe—which is what Cavanagh was interested in. Likewise, with the design of phenomenal consciousness. If different versions have proved equal in terms of survival, the details won't have mattered, and subpopulations of sentient animals could well have taken their own paths.

This has to mean that, as things stand today, it's possible—and even likely—that the phenomenal feel of the same sensory events could differ markedly between individuals. The philosopher John Locke famously raised the possibility of 'inverted colours' in his *Essay* of 1690:

> If the idea that a violet produced in one man's mind by his eyes were the same that a marigold produces in another man's, and vice versa...this could never be known: because one man's mind could not pass into another man's body, to perceive what appearances were produced.[64]

But we don't have to agree it could *never* be known. It could be known, at the level of brain activity, to a neuroscientist like Mary, who could indeed pass into another man's body. And it could presumably be known, at the level of behaviour, to anyone keen-eyed enough to take account of *everything* that happens next.[65] All the while, however, it could remain unknown to natural selection.[66]

I may say I rather like the implication: that there could be more diversity in the quality of sensations than, as evolutionary determinists, we might have guessed. Perhaps we really shouldn't assume that other people see red, or taste honey, or feel pain exactly the way we ourselves do. Furthermore, perhaps Mary shouldn't assume that her studies of other people will necessarily have revealed what it's going to be like *for her* to see red. What if Mary herself is an outlier, a variant that wasn't included in the population she's been studying? Then, she might learn something, after all, when she first sees red.

11

EVOLVING SENTIENCE

Today, sentience may—possibly—be all around us. But there was a time in the history of living organisms when it did not exist anywhere on Earth. Given that humans had non-sentient ancestors, There has to be a story to be told about how our ancestors got from there to here.

Evolution is not forward looking. Nonetheless, I think our strategy should be to engage in *forward engineering*. That means beginning with the end product, phenomenal consciousness as humans experience it today, but rather than treating this, as analytic brain science does, as something to *deconstruct*, we should treat it as something to *invent*. Even though it can't have been the goal of natural selection (natural selection doesn't have goals), we can make it *our* goal to come up with a story that will get us from insentience to full-blown sentience.

Since we're discussing evolution, we can assume three guiding principles. First, there must have been a continuous sequence of stages with no unaccountable gaps. Second, every stage must have been viable, at the time, on its own terms. Third, the transition from one stage to the next must always have been an upgrade, adding to the chances of biological survival.

As a model case of forward engineering, let's begin with a simpler example: the evolution of the human eye. If you take the eye as it is and try to deduce its history, you will—as many critics of evolution have pointed out—find it very hard to see how it could possibly have been assembled from scratch by natural selection. However, if you start with a patch of light-sensitive skin and aim to invent an eye, it turns out not to be so difficult after all.

Here goes. The patch of light-sensitive skin dimples to form a curved pit so that light coming from different directions produces different gradients of illumination. Next, this pit deepens further to become a spherical cavity with a small entrance hole, causing an image to form as with a pinhole camera. Next, transparent skin grows over the pinhole to protect the cavity from dirt. Next, the skin thickens to become a lens.

Charles Darwin confessed that he himself believed at one point that the suggestion that the eye evolved by natural selection might seem 'absurd in the highest degree'. But he was quick to answer his own doubts:

> If numerous gradations from a simple and imperfect eye to one complex and perfect can be shown to exist, each grade being useful to its possessor, as is certainly the case; if further, the eye ever varies and the variations be inherited, as is likewise certainly the case; and if such variations should be useful to any animal under changing conditions of life... then a perfect and complex eye could be formed by natural selection.[67]

Admittedly, when it comes to sentience, we have less to go on. If there are living examples of animals that are only part way to being fully sentient, we don't yet know how to recognize them. Moreover, we can't be sure that 'numerous gradations' do still exist. But, while Darwin himself always emphasized gradualism, his theory allows for there to be rapid stepwise changes. If a newly acquired improvement turns out, by a stroke of serendipity, to provide an unexpected springboard to an even better one, the intermediate stage will be short-lived.

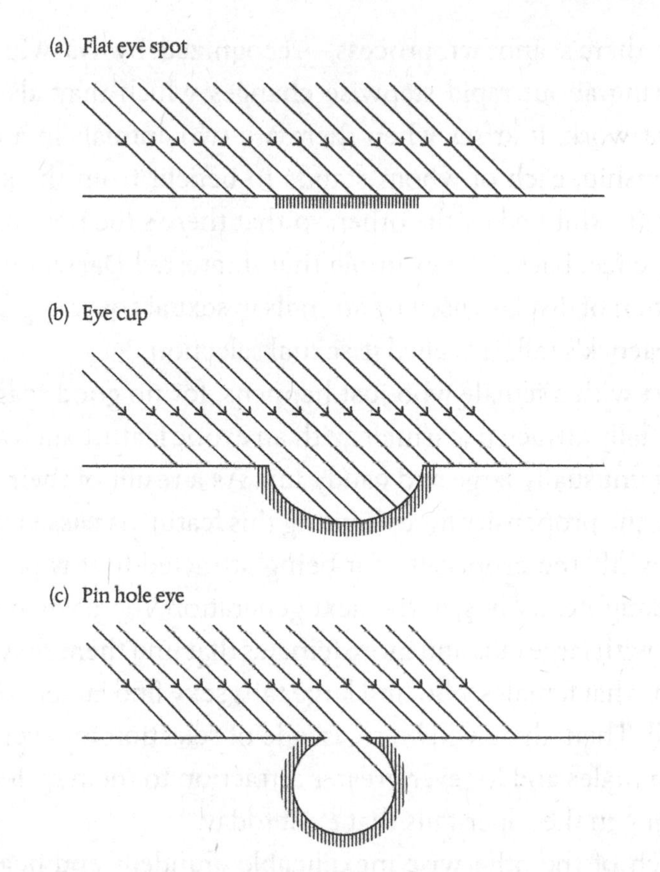

Figure 11.1 Stages in the early evolution of the eye

Indeed, just such a step seems to have occurred in the evolution of the eye. The transparent skin that first covered the hole of a pinhole-eye served to keep dirt out. But, as luck would have it, this skin could easily be built up to form a lens. So, once the skin covering developed, evolution was never going to rest there without exploiting its potential. That's why no living animal has plain skin across the pinhole. Among species of cephalopods, for example, Nautilus still has a completely open eye-hole, but all others, such as octopuses and squids, have a well-formed lens.

In the evolution of sentience, as we'll see shortly, there were indeed a series of lucky accidents. I think these may explain why it evolved quickly and why the intermediate stages may be missing.

But there's another process—recognized by Darwin—that can bring about rapid stepwise changes which may also have been at work. It arises when there are two animals in a dyadic relationship, each of whom stands to benefit from the success of a trait exhibited by the other, so that there's the possibility of positive feedback. The example that interested Darwin was the evolution of displays used by animals in sexual courtship such as the peacock's tail. He called it sexual selection.

Start with a female who just happens, for no good reason, to be sexually attracted to a male with an exotic feature such as having an unusually large and gaudy tail. As a result of their sexual union, the propensity for exhibiting this feature is passed to their sons, while the propensity for being attracted to it is passed to their daughters. Thus, in the next generation, there will be more males with large tails and more females that find them sexy. Now, assume that females who find large tails sexy find larger tails sexier still. Then, there will be a cascade of selection for ever larger tails in males and for ever greater attraction to them by females, resulting in the super-tails that exist today.

Much of the otherwise inexplicable grandeur and beauty of courtship displays can be attributed to sexual selection. Darwin believed it was responsible for humans' love of music, art, and poetry—traits that seem to be absurdly over the top when it comes to practical returns. Could the same be said of phenomenal consciousness: that it's unnecessarily wonderful? If so, could something like sexual selection, at the level of *mind mating*, have been responsible for making phenomenal properties the runaway success they have become?

We'll have to see. After all these preliminaries—the definitions, the arguments, the analogies—it's time to put my particular theory on the line.

12

THE ROAD TAKEN

We have arrived at a certain picture of sensations as humans experience them. Sensations are ideas that represent what's happening at your sense organs and how you feel about it. They do this by tracking motor responses that the stimuli evoke in your brain: covert—unrealized—forms of bodily expression. They have acquired phenomenal properties as a secondary overlay. These properties are not illusory. They are veridical properties of the feeling of 'what it's like'.

With human sensations as the end point of an evolutionary process, I now want—as I promised—to go back to the beginning and try to reinvent them. I'll spin it as a developing narrative. Figures 12.1 and 12.2 can serve as an illustrative crib.

Imagine a primitive amoeba-like animal floating in the ancient seas. Stuff happens. Light falls on the animal, objects bump into it, chemicals stick to it. Some of these surface events are going to signal an opportunity to be embraced, others a threat to be avoided. The animal, if it's to survive, must have evolved the ability to sort out the good from the bad and respond appropriately—with a wriggle of acceptance or rejection (Figure 12.1a). When salt arrives at its skin it detects it and 'wriggles saltily'. When red light falls on it, it makes a different kind of wriggle; it 'wriggles redly'.

The responses are automatic, reflex behaviours that reach out to the stimulus with an evaluative response. They will have been honed by natural selection to be precisely adapted to the stimulus events—taking account of their quality, intensity, and distribution on the body surface and the implications they have for the animal's well-being. To begin with, the responses are organized locally at the body surface. But before long, in order to allow coordination, sensory information gets to be sent to a central ganglion or proto-brain where a reflex response is initiated (Figure 12.1b).

Let's call these reflex responses 'sentition', something that hovers between sensation and action. Sentition *enacts* what the stimulation *means* to the animal. Indeed, it makes the meaning public. So an outside observer, if there were to be one, could tell from what the animal *is doing* just how it *feels* about what's happening. Yet, at this early stage, the animal itself doesn't create any kind of mental picture of what's happening to it and has no feelings of its own.

However, as these animals evolve and begin to lead more complex lives, the time comes when reflex behaviour is not enough. If they are to behave more flexibly, they need to be able to store information about themselves and their environs in a form they can refer to offline. In particular, they need a way of representing and holding 'in mind' information about events occurring at their body surface. But how are they to move to this new level?

It so happens that there's an ingenious way of doing it, based on sentition. In the same way that an outside observer could tell what the animal is feeling from what it's doing, so, in principle, could an *inside observer*. In other words, the animal will be able to discover for itself what the stimulation means to it *by monitoring its own response*. And there's a simple trick for doing this. When its brain sends motor commands to create the response, all it has to

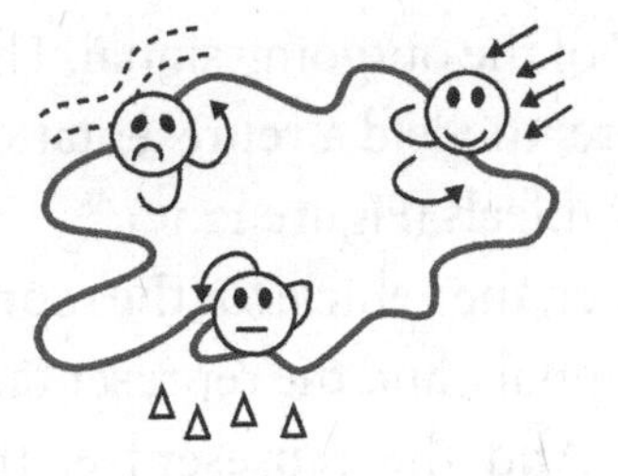

(a) Evaluative response occurs at site of stimulation

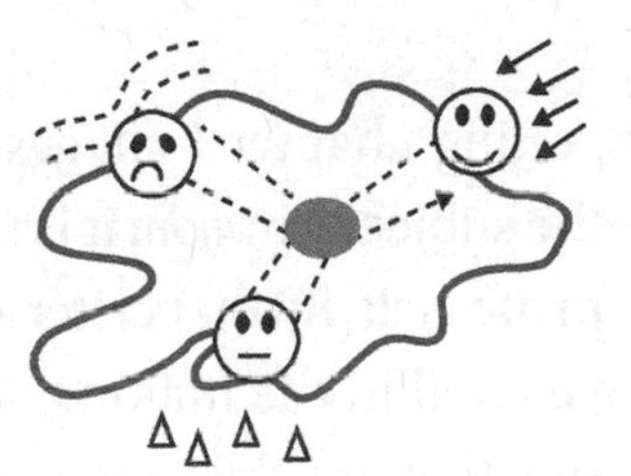

(b) Response comes under central control

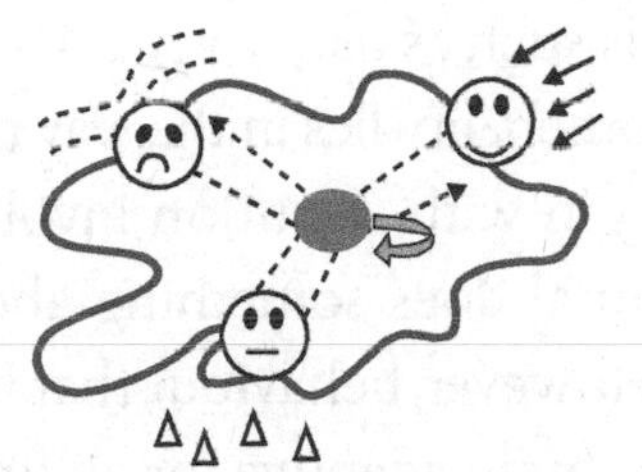

(c) Motor command signal is copied, to represent what the stimulus feels like

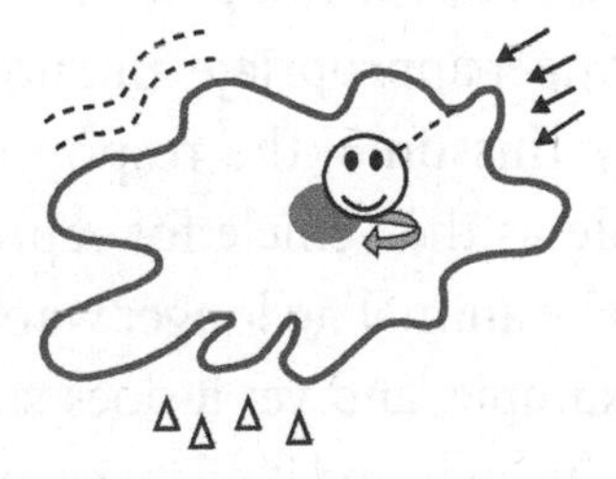

(d) Response becomes privatised—efferent copy is retained

Figure 12.1

do is make a copy: an 'efference copy' of the outgoing signal. This then can be read, in reverse, as it were, to yield a representation of how it's responding and so of how it feels (Figure 12.1c).[68]

In the language we established earlier, the vehicle for the representation is a copy of the command signals, and the representandum is the occurrent stimulation. And the representee, the subject the representation is for? Can we say that once an animal begins to represent its situation this way, it's on the way to having a *self* that is the *subject of sensations*?

You may remember that in the opening chapter I suggested that the subject of any mental state—the subject *for whom* it is that state—should be counted as at least a proto-self. By that criterion, this primitive animal does indeed have a self in the making, and this self is the subject of representations that are the precursors of sensations as we humans know them. Yet, of course, sensations at this stage don't have any of the remarkable phenomenal feel that they have in sentient animals such as us.

The key to the emergence of phenomenality lies in the way that sentition continues to evolve. To begin with, sentition involves an overt adaptive response: the animal does something about what it senses to be happening to it. However, behaviour that was adaptive at the start is not going to remain adaptive for all time. As these animals develop more sophisticated ways of interacting with the environment, there is bound to come a point when the original bodily responses are no longer appropriate. But now there's potentially a problem: for, by this time, the responses have already acquired their useful role as the vehicle for representing what the stimulation means. The animal no longer wants to recoil reflexly from red light, for example, and yet it does still want to know that red light is falling on its body and it's dangerous.

Then, what to do? The answer is, again, ingenious. It is for the responses to become internalized or 'privatized' (Figure 12.1d).

The outgoing commands, rather than causing an actual bodily response to the stimulus where it's occurring, begin to target the internal body map where the sense organs first project to the brain. This way, the commands can retain their crucial intentional content: they are still about responding to what's happening to *me* with *this* part of *my* body. There is still an efference copy, forming a representation the subject can milk for information. But now the commands issue in a virtual, as-if, expressive response that no longer shows on the surface.

So, here comes the next break of fortune. (The story continues in Figure 12.2, where it's updated to a modern-looking brain.) Once motor signals that were formerly sent out to produce a response at a particular locus on the body surface have been redirected to the place in the brain where sensory signals from this locus come in, there's the potential for *feedback*. When conditions are ripe, the outgoing motor signals will be able to interact with the incoming sensory signals to create a self-entangling loop—a loop that can sustain recursive activity, flowing round and round, catching its own tail (Figure 12.2c).

Remember how, in my experiments recording from cells in the monkey's brain, a loop got established accidentally between what the cell could *hear* and the *sound* its own electrical response produced on the loudspeaker—leading to that self-sustaining Whoosh!? Now something like this begins to occur with sentition; and the consequences are potentially game-changing.

It means, to start with, that sentition can be drawn out in time, so that the subject, monitoring the outgoing signals, will get the impression that each moment of sensation lasts longer than it really does. Sensations are, as it were, being thickened up. But this is only the beginning of a more elaborate transformation. Once the loop has been established, the circling activity can be channelled

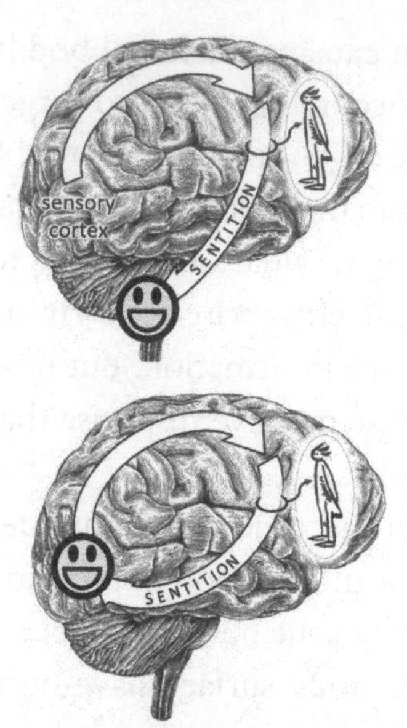

(a) *Sensation*. A module in the brain—a proto-self—forms a mental representation of what the stimulation *feels like* by monitoring the response.

(b) *Privatization*. The response becomes internalized so as to address the body map where sensory signals arrive in the brain.

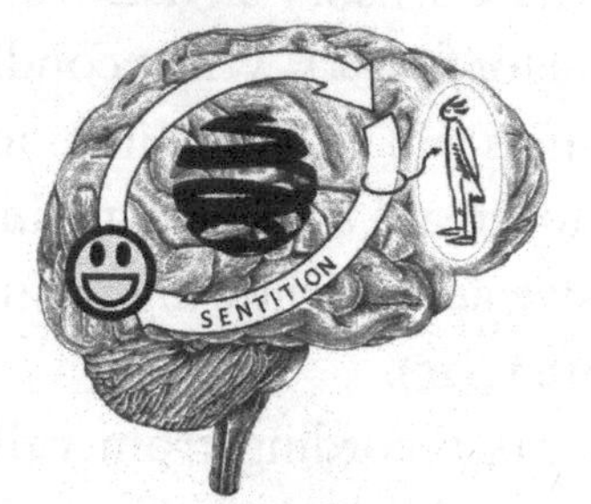

(c) The *thick moment*. A feedback loop is created between sensory input and motor response so the activity becomes recursive and stretched out in time.

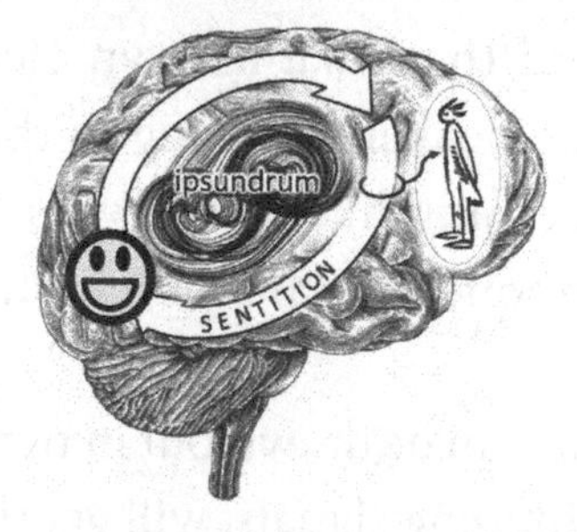

(d) The *ipsundrum*. The recursive activity settles into an attractor state.

Figure 12.2

and stabilized so that it settles into an 'attractor' state, where a complex pattern repeats itself over and over (Figure 12.2d).[69]

Such an attractor, while being a real mathematical object, can have far-out, almost unthinkable, properties. Indeed, from here on, whenever the opportunity arises to 'improve' the quality of sensations, natural selection has a whole new design space to explore. Small tweaks to the circuitry can have dramatic effects on the subject's reading of what sensations feel like. The upshot is that sensations come to be experienced as being inalienably private, suffused with distinctive modality-specific qualities, rooted in the thick time of the subjective present, made of immaterial mind stuff: in short, phenomenal.

The kind of mathematical attractor we're talking about deserves a name. I've called it the 'ipsundrum', a self-generated conundrum, echoing the name of the real impossible triangle, the Gregundrum. I like this name because it makes the vehicle for phenomenal sensations sound not only peculiar but also substantial and thing-like—which, of course, it must be, even if only as a mathematical object. The ipsundrum is something about which we can ask, down the line: is the brain of this animal (or this robot) capable of constructing and reading such a thing?

And if the name stops you in your tracks, that's as it should be. For there's no question that the invention of the ipsundrum is a shocking evolutionary development. Out of nothing, apparently flying in the face of the principle of sufficient causation, natural selection has created a piece of magic and planted it in the brains of billions of sentient animals like ourselves.

What happens next? At a psychological level, there's a major consequence: a step-change in how sensations contribute to the *sense of self*. Sensations have always been essentially *personal*. For you, the subject, they represent *your* interest in the stimulation at *your body surface*, and they do so by taking account of what *you* are doing in response. It means that in the act of representing 'what's happening to me', you are filling out your sense of 'what I am'. But if, now, as a result of these developments, your picture of 'what's happening to me' comes to seem ever more impressive and exotic, then so must be your idea of what 'I' am. A self that was not much of a self is being lifted, all of a sudden, to the status of a phenomenal self, a self worth having.

And I mean all of a sudden. The invention of phenomenality and the consequent elevation of the self will likely have come about very rapidly. The reason is that self-sustaining feedback of the kind that sustains the ipsundrum is *all-or-nothing*. Depending on the precise coupling between input and output in the feedback loop, the activity will either take off and outlast the stimulus or it won't (think of what happens when a microphone gets close to a speaker in a lecture hall). There is an abrupt transition from one state to the other: the species of animal that was insentient wakes up to find itself living in the thick moment of phenomenal consciousness.

In sum, I believe this is a trajectory that evolution *could* have followed, all the way from primitive wriggles to full-blown phenomenal sensations. In fact, I dare say it's the trajectory that evolution *was destined* to follow because of the existence of those lucky breaks: points at which one evolutionary stage unexpectedly provided a springboard to the next.

This happened at least three times. (a) Command signals for the reflex response to sensory stimulation could be exploited to represent what the stimulation meant. (b) The requirement to privatize these responses created the conditions for sensorimotor feedback loops. (c) These loops had the potential to create attractors, that could represent surpassingly strange properties.

Without these strokes of serendipity, the evolution of phenomenal consciousness would have stalled. It is, indeed, *lucky* that evolution could take this route. However, this isn't to say it was improbable. In fact, the opposite is true. For, as we've seen in our reinvention, the opportunities were there for the taking and always would have been. That means if history were to be re-run, evolution would very likely take the same path again.[70]

13

THE PHENOMENAL SELF

We've been assuming that the destination of the evolutionary story is phenomenal consciousness and the enhancement of the self. It's time to look more closely at how these go together and what the pay-off is likely to have been. I've referred to the phenomenal self as a 'self worth having'. But it remains to be seen why a self that's worth having subjectively is worth having in terms of biological survival.

The self we're talking about is the Cartesian self: the self that you discover by introspection to be the subject of your mental states—your 'I'. Descartes had a method for discovering the nature of this self. It was to engage in forensic self-examination, disregarding everything he could reasonably argue was inessential until he reached bedrock: the properties his 'I' must have in order to exist at all. As a result, he famously concluded that the one thing he couldn't doubt to be essential were his thoughts. *I think, therefore I am.*

It's hard to quarrel with the method. But many did and do disagree with the conclusion. The philosopher David Hume arrived at an end point much closer to where our own discussion has been heading. He decided *I feel, therefore I am:*

> For my part, when I enter most intimately into what I call *myself*, I always stumble on some particular perception or other, of heat or cold, light or shade, love or hatred, pain or pleasure. I never can catch *myself* at any time without a perception, and never can observe any thing but the perception. When my perceptions are remov'd for any time, as by sound sleep, so long am I insensible of *myself*, and may truly be said not to exist.[71]

Writing fifty years before his fellow Scotsman, Thomas Reid, Hume did not distinguish between sensation and perception and used perception as the generic term for feelings of all kinds. But his meaning is plain. 'When I am insensible of myself I cease to exist.' What's at issue is sensory phenomenology.

In the absence of sensations, 'I' am not. 'Touch me mother, that I may be here.'

Hume did not, however, immediately embrace this insight as the profound truth that I think it is. He considered that the self he had discovered lacked *character*, that there was nothing to bind its elements together to make a significant whole. '[We] are nothing but a bundle or collection of different perceptions, which succeed each other with an inconceivable rapidity, and are in a perpetual flux and movement . . . nor is there any single power of the soul, which remains unalterably the same, perhaps for one moment.' He likened the mind to a theatre, with a constantly shifting display of sensations that 'successively make their appearance; pass, repass, glide away, and mingle in an infinite variety of postures and situations'. He went on, 'There is properly no *simplicity* in it at one time, nor *identity* in different . . . nor have we the most distant notion of the place where these scenes are represented, or of the materials of which it is composed.'

I believe Hume called this quite wrong. It's true that sensations can seem to be disturbingly evanescent—no sooner come than gone, 'a moment white, then melts forever'. There is, however, a feature that more than makes up for this and provides the self with the simplicity and identity Hume thought was lacking. It is that your sensations are *intrinsically personal*. They are viewable to you and no one else, they relate to events at your own body, they are located in a space and time of your own making, and they have qualities—phenomenal redness, saltiness, paininess—that are entirely your own invention. Every sensation bears your unique phenomenal signature. In fact, for all you know, you may be the only 'I' in the universe for whom seeing red, tasting salt, touching a nettle feels *like this*.

These authorial markers, common to one sensation and the next, are more than sufficient to link each episode of your 'I' to the immediately preceding one and so to establish your self's continuing existence—even when, as Hume puts it, your sensations are removed for a time, as by sound sleep. You need never doubt it's the same 'I' that picks up where it left off because it is undoubtedly doing sensations *your way*.

*

Imagine that your sensations form a sequence of paintings that hang in a long picture gallery stretching back into your past. Their subject matter is what's been happening at your sense organs and how you've felt about it at successive moments. The style in which they are executed is peculiarly yours. Just as we say paintings by Picasso are all 'Picassos' or those by Cézanne are all 'Cézannes', these are all 'Yous'.

Note that these 'Yous' are not mere copies of the facts of sensory stimulation; rather, they represent your creative take on it. In this, they resemble the work of proper artists. In Paul Klee's words, 'Art does not reproduce the visible; rather, it makes visible.'[72] In Pablo Picasso's, 'Art is a lie that makes us realize truth... Through art we express our conception of what nature is not.'[73] For Eugene

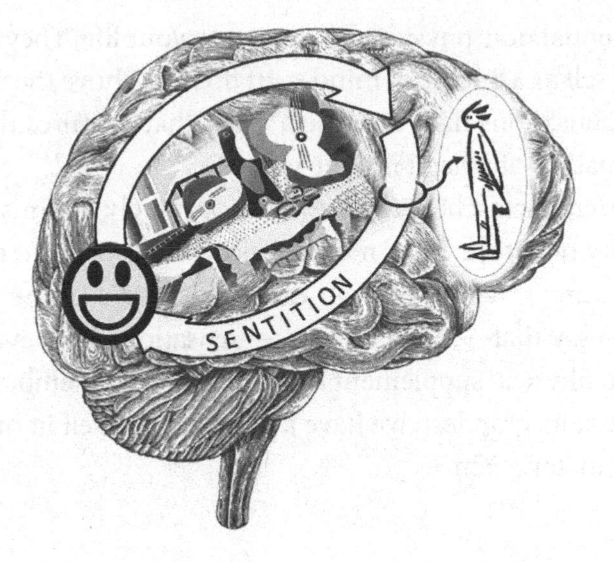

Figure 13.1 Consciousness as art

Delacroix, 'The subject [of the paintings] is yourself; they are your impressions, your emotions before nature.'[74] For Vincent Van Gogh, 'What I most want to do is to make of these incorrectnesses, deviations, remodellings, or adjustments of reality something that may be "untrue" but is at the same time more true than literal truth.'[75] Samuel Palmer wrote in his notebook, 'Bits of nature are generally much improved by being received into the soul.'[76]

Would it be going too far to borrow these words and say that sensations make visible to you what is invisible to everyone else: the strange truth about your bodily presence in nature? Maybe it would. But however we say it, I want to make a serious scientific point. Sensations that far back in history started out as a way of tracking your interaction with the physical world have, in the course of evolution, come to play a subversive double role. While still connecting you to the physical environment, they now serve also to distance you from it. They give you the feeling that there is an essential non-physical dimension to your life. They fix your sense of self as a bubble of mind stuff floating above the world of matter. Sensations have become a work that captures the paradoxical nature of what it is to be you.

Friedrich Nietzsche wrote, 'Art is not merely an imitation of the reality of nature, but a metaphysical supplement to the reality of nature... We have art in order not to die of the truth.'[77] I want to say that—for us humans—sensations have evolved to be a metaphysical supplement to the reality of our embodiment. To put it at its grandest: we have a phenomenal self in order not to die of materialism.

*

Why will natural selection have gone for this? There's an answer for humans that is as obvious as it is surprising. In the course of relatively recent history, those of our ancestors who thought of themselves as beings imbued with immaterial qualities, existing outside normal space and time, will have taken their own existence ever more seriously. The more mysterious and unworldly the qualities of phenomenal consciousness, the more significant the self. And the more significant the self, the greater the value that people will have placed on their own—and others'—lives.

Once humans developed the capacity for language, they no doubt began to discuss these weighty issues between themselves. It can't have been long before they parlayed the phenomenal self into the culturally enriched idea of the 'soul'—a self that transcends physical embodiment and continues to have needs and ghostly powers even beyond death.

But that's humans. And it can hardly have been at this airy level that the phenomenal self first proved its worth in terms of biological survival. So, how does it work for those animals who are phenomenally conscious but never get to talk about the self—who have an idea of 'I' but not a word for it? Can they think of others as having an 'I' like theirs? Will they be able to exploit this to their advantage?

As I'll show when we come to specific examples later in the book, there's evidence that some non-human animals do indeed see other animals as individuals who have minds of their own with a unique 'personal' identity. In other words, they see others as conscious *subjects*, not simply as physical *objects*, as *beings* not simply as *bodies*.

Even for an animal without language, this kind of respect for the selves of others can make all the difference to how you manage your affairs. When you see another individual—a mate, a mother, a friend, an enemy—as having a self like yours, you'll have a head start in understanding them and predicting their behaviour. Not only will you be able to give more thought to the other's needs and capacities, but you may even be able to take account of *their* thoughts about *you*. The fact is you'll now be on the road to becoming a 'natural psychologist'. If you can imagine yourself in someone else's place, you can model *their mind* on your own. What's more, you'll find that your ability to mind-read is greatly facilitated by the fact that your own mind—thanks to having phenomenal sensations at its core—comes across to you as having an easy-to-grasp internal structure.

This possibility, that the great benefit of phenomenal consciousness is that it provides you with a 'mind-fable' or even a 'book of the mind for dummies'—has been gaining traction with philosophers in recent years. Dennett has long argued that introspection provides you not with a picture of your brain as such (which, of course, you wouldn't understand) but with a semi-fictional narrative. 'We don't see, or hear, or feel, the complicated neural machinery churning away in our brains but have to settle for an interpreted, digested version, a user-illusion that is so familiar to us that we take it not just for reality but also for the most indubitable and intimately known reality of all.'[78] For Michael Graziano, 'our intrinsic model of a mind is a cartoonish version of the functioning of an active, attentive brain'.[79] For Keith Frankish, 'representations of phenomenal properties are simplified, schematic representations of the underlying reality'.[80] And for David Chalmers, '[introspection] needs to keep track of similarities and differences in mental states, but doing so directly would be inefficient, and it does not have access to underlying physical states. So it introduces a novel representational system that encodes mental states as having special qualities.'[81]

I agree. But this is all rather hand-wavy, and the philosophical discussion is short on examples of how it might actually work. So, here, let me offer one. It's one of the simplest cases of mind-reading: how you recognize in someone else which sense organ they are employing to find things out.

*

I assume it's as obvious a fact as any in your experience that each sensory modality has its separate and peculiar quality space. All sensations mediated by your eyes are distinctly visual (and, as we saw with phosphenes, still visual, even if the stimulus is touch rather than light), all sensations mediated by your ears distinctly auditory, and so on. Within a single modality, sensations are part of a continuous spectrum, but between modalities there's an unbridgeable gap. You can imagine, for example, one colour sensation changing to another colour, one sound to another sound, one smell to another smell; but you can't get from a colour to a sound or from a sound to a smell. The distinctness of the sensory modalities is so evident and absolute it may seem to be a basic fact of nature, as if the modalities are 'natural kinds'—groupings that reflect the deep structure of the natural world.

Yet, in truth, there are no such distinct kinds to be found either in the environment or in the physiology of the brain. The receptors in different sense organs have all evolved from a single kind of hair-like cell, the sensory *cilium*, and they respond electrochemically in just the same way. The information they send forward is carried by impulses in the same kind of nerves. And, though the information then undergoes modality-specific processing, to deliver representations that typically describe different kinds of objects, there's no stage at which a qualitative gulf appears. A physiologist, looking at nerve cells in the incoming pathways of the brain, couldn't tell from the pattern of firing whether the cell was signalling information about light or sound or touch.

Despite this, when scientists make maps of the brain, they often colour-code the sensory pathways. Thus, for example, the legend to a recent state-of-the-art brain map reads: 'the image above shows areas connected to the three main senses—hearing (red), touch (green), vision (blue)'. The purpose here, presumably, is to distinguish pathways that have different *functions for the user of the brain*. By analogy, a map of the London Tube shows the lines in different colours: Bakerloo line brown, Central line red, and so on. The purpose is to distinguish lines whose trains take people to different places. But with the sensory pathways, the question is: just what are their different functions? If they take people to different places, what places are these?

We can't answer such questions at the level of the brain. We need to pose them at the level of behavioural ecology. How, in practice, do you use your different sense organs in negotiating the world you live in? The answer then becomes obvious. Your different sense organs sample stimuli from different parts of the external world. And most important, they open windows onto sub-worlds that have quite different *affordances*—opportunities for action.

Think about which of the senses you rely on when making decisions about, respectively: eating, climbing, picking, kicking, caressing, swatting, scratching. Your sense organs clearly have different zones of relevance. Note how you can *see* fruits in a tree but not hear them, you can *feel* the temperature of a lake but not smell it, you can *hear* a spoken conversation but not taste it, and so on.

This is so familiar that I'm sure you hardly stop to consider it. However, it has major implications for mind-reading. It means that if you are trying to predict another person's behaviour by imagining yourself in their place, one of the first things you should do is to home in on what sensory modality they are using—get on their wavelength, as it were. And this is precisely where the phenomenal quality of your own experience will help you, for it means you have a ready-made set of filters to apply. As you imagine what it's like to be the other person, your thoughts are channelled into the appropriate sensory zone. Thus, you narrow the range of behaviours to expect.

If this seems obvious, that's exactly the point. You wouldn't think of doing it any other way. Qualia-coding the sensory modalities, if we may call it that, seems like second nature. Yet, to say it again, qualia-coding actually goes beyond nature: it introduces an absolute difference of kind between ways of gathering sensory information from the environment that are physiologically on a continuum. Your mind is, as it were, taking 'artistic licence' with how it represents what's going on in your brain. As Picasso might have put it, qualia-coding is 'a lie that helps you realize truth'.

*

Let me broaden this out. In the 1980s, having introduced the notion of humans—and gorillas—as natural psychologists, I suggested that mind-reading was made possible by the evolution of an introspective organ—the 'inner eye'—that looks in on the workings of the brain. In a lecture at the American Museum of Natural History,[82] I explained:

> The inner eye provides a picture of its information field that has been designed by natural selection to be a useful one—a user-friendly description, designed to tell the subject as much as he requires to know in a form that he is predisposed to understand. We can assume that throughout a long history of evolution all sorts of different ways of describing the brain's activity have in fact been experimented with—including quite possibly a straightforward physiological description in terms of nerve cells, RNA etc. What has happened, however, is that only those descriptions most suited to doing [natural] psychology have been preserved. Thus the particular picture of our inner selves that human beings do in fact now have—the picture we know as 'us', and cannot imagine being of any different kind—is . . . the description of the brain that has proved most suited to our needs as social beings . . . Consciousness is a socio-biological product—in the best sense of socio and biological.

In that same lecture, I compared mind-reading to a kind of mind transplant and emphasized how, as with any other organ transplant, it's important that the receiver and donor are compatible. That's to say, if you are to understand other people by putting yourself in their place, you must be able to assume that their minds and yours run on similar principles; better still, that you each picture your minds in the same way.

For humans, this is in fact continually tested and brought back on track by the revolving dynamics of interpersonal encounters. As your life unfolds, it becomes ever more apparent to you how your understanding of yourself helps you understand other people—and, for that matter, how it helps you to be understood by others.

The mutual adjustment and refinement of mental models that flows from this will be a work in progress that continues throughout your individual life. As a human, you will have language and culture to support you. But, long before humans became human, natural selection will have seen to it that the basic structure of phenomenal experience, with which you are born, is the structure with which others of your species are born too.

I believe the importance of *mind-reading in both directions* may help explain features of phenomenal consciousness that would otherwise seem superfluous. Earlier, I called attention to 'sexual selection' and how it can produce positive feedback, leading to a runaway beautification of animal courtship displays. The more attractive the tail of a peacock to a peahen, the greater the advantage to her male offspring in having a tail that's even more attractive. Thus, the tail becomes in effect self-selecting. I raised the question of whether anything like this could have occurred with phenomenal consciousness. Perhaps we see now just how such positive feedback might have happened. The more phenomenal consciousness helps the mind donor with the task of mindreading, the more it helps the recipient with the task of being mind-read. Thus, phenomenal properties could indeed have become self-selecting, leading to their spiralling up into the phenomenosphere, to the heights of strangeness and beauty we enjoy today.

Geoffrey Miller, author of *The Mating Mind,* has argued explicitly that consciousness has not just a social function but also a sexual one. Given that mind-reading between mates—or potential mates—is probably the arena in which the skills of a natural psychologist matter most of all, I'd say he's onto something.

14

THEORETICAL MISPRISIONS

There's an objection to this evolutionary story that may have occurred to you; at any rate, it has occurred to others who have commented on it. It's that, even if this story could in principle explain how phenomenal consciousness has evolved and had survival benefits, it doesn't establish that *only* phenomenal consciousness could have played this role. I've been arguing that things evolved the way they did because natural selection found an opportunity—by creating sentience—to improve survival prospects for social creatures who value themselves as individuals. I've argued that the solution selection found was *sufficient* for the purpose. However, I certainly haven't shown that this solution was the *only* way of doing it.

This has led the psychologist Stuart Sutherland to write:

> There is, unfortunately, an obvious fallacy in the argument. The brain could represent the processes underlying motives, thinking and so on and could use this representation as a model for others' behaviour and the forces underlying it without the representation appearing in consciousness...It is easy to invent functions that consciousness might subserve: what is

> difficult and has never yet been achieved is to show that they can be subserved only by consciousness.[83]

Of course, he's partly right. At the start of this book, I conceded that a creature that is cognitively conscious but lacks phenomenal experience might nonetheless be able to develop some sort of self-concept and even acquire a theory of mind. Phenomenal consciousness is not a logical requirement for selfhood or mind-reading. In fact, when human engineers get to building advanced social robots, they may find ways of doing it, satisfactorily, that don't require sentience.

Maybe so. However, I'd answer Sutherland by pointing out that the theoretical possibility of doing things differently doesn't mean there was a practical possibility of biological evolution taking a different course. Even if an artificial brain could represent the mind in a way that doesn't involve phenomenal consciousness, it's by no means clear that *our* brains could have evolved to do this or that, if they had, this would have provided us with a working model of the mind anything like as useful and user-friendly as the one we've ended up with. My argument is not that mind-reading couldn't in principle be subserved by anything other than phenomenal consciousness, it's that it can in fact be subserved particularly effectively by it, and that—as luck would have it—an evolutionary pathway to it was available.

David Chalmers has raised a similar kind of objection to my evolutionary account, an objection of the form: 'Yes, but why was it necessary for natural selection to solve the problem *that* way?' Humphrey, he writes:

> suggests that, in an evolutionary context, thinking of ourselves as conscious in this mysterious way makes ourselves seem more significant, which leads us to place more value on our and others' lives . . . This is an interesting idea . . . [Yet] most creatures

don't seem to have too much trouble placing a high value on their own lives to start with.[84]

Again, he's partly right. He's saying, justifiably enough, that a creature can perfectly well value its life for other reasons. The desire to stay alive is not something that can be motivated *only* by phenomenal consciousness. In fact, with many animals we can assume that a much more basic unconscious instinct for survival suffices. So, it might seem the solution I've suggested wasn't necessary. Given that our ancestors already had a life instinct, why didn't natural selection leave well alone? If it wasn't broken, why fix it in the exotic way for which I've been arguing? Other critics have raised the same issue. They say I seem to be suggesting an answer to a non-problem.

My response is that 'leaving well alone' is not how evolution works. Natural selection continually searches out opportunities for animals to increase their biological fitness by adopting—and adapting to—ways of life that were previously out of reach. In fact, that's why there's an arrow to evolution: why animals move on into new environments of adaptiveness, even when they were already doing quite nicely as they were.

Birds, for example, evolved wings and took to the air even though life on land was—and remains—quite satisfactory for those who stayed behind. Would anyone say that wings were the answer to a non-problem? No, they were the answer to what must have *become* a problem for the ancestors of birds that sought to enter the ecological niche that would be opened up by becoming airborne: namely, how to defy gravity and stay aloft. They didn't *have to* move into this niche but, for those that did, wings became their passport to it. Likewise, I'd say, phenomenal consciousness was the answer to what must have *become* a problem for the ancestors of sentient creatures that sought to enter the niche that would be opened up by becoming natural psychologists: namely, how to develop an individualized sense of self and the ability to mind-read. They didn't *have to* move into this niche but, for those that did, the phenomenal self made it accessible.

But this brings us to a different kind of objection that has been raised by critics. They ask: not why was it needed, but why isn't there more of it? Given that phenomenal consciousness, centred on sensations, does such a good job of amplifying the sense of self and providing a picture of how the mind works, why was the phenomenalization of mental states not taken still further? Why didn't natural selection go one better and arrange for other mental states—beliefs, perceptions, intentions, and so on—to have their own characteristic phenomenal feel? As Chalmers says, 'it is not really clear why access to a [sensory] modality as opposed to an attitude should make such a striking difference'.[85]

I can't but agree that, if other kinds of mental state were to have their own phenomenal signature, this might possibly make mind-reading even easier and the phenomenal self still more worth having. Therefore, the fact that it's only sensations that have acquired phenomenal properties calls for explanation.

Luckily we have an answer ready and waiting. Sensations, according to the theory, originated as evaluative responses to sensory stimulation, a form of bodily expression that the subject reads to get a picture of what's happening. When this response is privatized, it creates the potential for feedback loops that can be elaborated to create the complex attractors that underwrite phenomenal experience. But note how this particular *history* is crucial. It's precisely because sensations originated in bodily expression that they could go on to acquire phenomenal properties; and it's because other mental states did not originate this way that they could not.

No surprise, then, that sensations have come to play such a unique role in grounding the sense of self. You feel, therefore you are. However, natural selection never had a chance to side with Descartes. You can *think* as long as you like, and you won't *be there.*

15

COMING TO BE: SENTIENCE AND BODY SENSE

If thoughts can't ground the self, the same goes for perception. When we discussed the case of H. D., the woman whose eyesight had been partly restored but whose visual cortex was apparently no longer functioning, I suggested that when visual perception exists in the absence of visual sensation, the experience of seeing doesn't bring with it a feeling of subjective presence. It's for this reason that H. D. found it so disappointing.

Such a dissociation between sensation and perception is, of course, quite unusual. It's not something most of us would presume to know about first-hand. Yet, there are examples of ordinary human experience that might make the dissociation not so foreign to us as we may suppose. I'm going to divert, for a moment, to highlight two areas of normal sensory experience where there is in fact an extreme imbalance between sensation and perception. One is 'bodily position sense', the other is 'orgasmic sense'.

*

Position sense, or proprioception, uses information picked up by sensors in joints and muscles to enable you to perceive the spatial location of your body parts. Your brain uses information relayed by proprioceptors to represent an objective state of affairs: for example, where the thumb of your left hand is located in space. Thus, in the dark, you can perceive where your thumb is, using proprioception, and in the light, you can perceive the same fact using visual perception—and these two perceptual representations, mediated by different sense organs, will agree.

However, what's notable about position sense compared to senses such as vision and hearing is that you don't have any accompanying sensation. Your brain is using the information passed on by proprioceptors but it's not providing you with a representation of the sensory stimuli as such. And, there being no sensation, there's no phenomenal dimension and no modality-specific quality. There's nothing *it's like*—in the dark—for you to have your thumb located where it is.

Position sense is, in fact, very like blindsight. It's a case of pure perceptual knowledge. If you were to be asked how you know where your thumb is, you'd find it a perplexing question; you might even have to admit that you were just guessing.

But now let's contrast the case of orgasm. Here it's all about sensation and very little about perception. The brain responds to incoming signals about genital stimulation by sending signals back down to the body with certain instructions—lubricate the vagina, stiffen the penis, pump blood harder, breathe faster. The intensity builds to a crescendo until it's released in an explosive rush. The heart rate doubles. In women, the uterus contracts rhythmically. In men, sperm-carrying semen is propelled out of the body. Your brain represents what all this feels like by monitoring its own command signals for these motor responses, as stretched out by feedback loops—telling you where it's occurring (starting in your genitals but spreading out), the timing (perhaps coming in waves), the phenomenal quality (the modality is akin to pain), and especially how good it feels.

However, what's notable here is that perception is largely absent. Orgasm is about your experience of the bodily events as such and very little about the objective external circumstances that are responsible for bringing them on. It's a case of pure what-it's-likeness. If you were asked how you know what orgasm feels like, you'd find the question perplexing for a different reason. It's simply obvious. How could you not know? No guessing needed.

Position sense and orgasmic sense are clearly outliers among human sensory systems. But their contrasting profiles fit nicely with the what and the why of the evolutionary story we've arrived at.

Position sense is what it is because proprioceptors have never been involved in evaluating and responding to *exogenous* stimulation. This means that, historically, there has never been a call for you to form a mental representation of 'the stimuli arriving at my muscles and joints and how I feel about it' by reading 'what I'm doing about it' (which will usually be nothing). By contrast, orgasmic experience is what it is because the primary role of sensory receptors in the genitals is precisely to elicit an evaluative response to stimulation from outside (and especially from another person's body). It is, we could say, an inter-body sense. And here the call has always been to form a representation of 'the stimuli arriving at my penis or my vagina and how I feel about it' by reading 'what I'm doing about it' (which now will certainly be something). In fact, orgasm could almost count as a poster boy for the general theory we've come up with of how sensations originated with responses targeted on the site of stimulation. We even speak of orgasm in active terms as 'coming' (and, in this case, a form of coming that has not been altogether privatized).

The upshot is that position sense, because it has no phenomenal quality associated with it, is of little or no importance to establishing your sense of self. While it gives you perceptual knowledge of how your body is positioned in space, paradoxically, it doesn't give you the feeling of *being there* in any deeper sense. An interesting consequence is that your idea of how your self relates to your physical body can be surprisingly labile. Recent research has shown that it's relatively easy to manipulate people's experience so that they have the illusion that the body that belongs to them is displaced from their real body and resides in the body of a model. More surprising still, people can even be persuaded that they possess a third arm extending from their chest. There's no sensation to either confirm or contradict it; it wouldn't *feel* any different if it were so.

Orgasm, by contrast, brings your sense of self sharply into focus. In this respect, orgasm is indeed a close cousin of pain. To quote climber Joe Simpson again, 'As the [searing agony] increased so the sense of living became fact.' Milan Kundera wrote in his novel, *The Unbearable Lightness of Being*, 'I think, therefore I am is the statement of an intellectual who underrates toothaches... The basis of the self is not thought but suffering, which is the most fundamental of all feelings. While it suffers, not even a cat can doubt its unique and uninterchangeable self.'[86] Could it be said, with equal justice, 'When it masturbates, not even a bonobo can doubt its unique and uninterchangeable self'? As we'll see later, it's not a silly question.

16

SENTIENCE ALL THE WAY DOWN?

Our discussion has centred on human beings. Human experience is bound to be the example that interests us most and that stands in most urgent need of explanation. But our own experience is also what incentivizes us to go further and ask how far the capacity for sentience extends beyond ourselves. It's because we are so impressed with what it's like for *us* that we urgently want to know what it's like for *them*: for other living creatures and perhaps machines.

Since we shall never be able sample other creatures' experience directly, we can only infer it from external evidence—from clues, such as there are, in their brains, behaviour, and natural history. In the absence of a theory, we'd be unable to read those clues; we wouldn't even know they *are* clues. Now we have a theory about the how and why of human sentience, we can begin the enquiry into sentience in other creatures by asking two leading questions about the external conditions that would allow it.

First, does the creature have the right kind of brain to deliver it? That's to say, a brain that, building on reverberatory sensory-motor loops, could generate attractors of the kind that we've identified as underlying human sentience? Next, does it have the right kind of lifestyle to require it? That's to say, the kind of life for which an elaborated 'sense of self' would enhance its personal and social survival?

In this way, we can use the theory to *exclude* a range of creatures from consideration. Then, for those who pass these two preliminary tests, we can proceed to ask some more specific diagnostic questions to decide who we should *include*. Does the candidate behave in ways that would be highly improbable in the absence of phenomenal consciousness? For example, does it show evidence in the lab or the field that it has an individualized sense of self that it prizes and that it extends this notion to other creatures of its kind?

So, a two-step plan. First, decide who couldn't possibly be sentient, using exclusion criteria. Then, apply more stringent inclusion criteria to decide who most probably is.

However, for this to work, we have to make a leap of scientific faith. We have to be confident our theory covers *every kind of sentience there is*. Suppose that, unknown to the theory, phenomenal consciousness can exist in forms not yet considered. What if it can be generated by an entirely different kind of brain process or bring different psychological benefits from those we've described? Then, our criteria for deciding who's in and who's out will be too strict. What if, rather than being an 'on' or 'off' feature, as our theory suggests, it can be present in different creatures along a continuum from less to more? Then, the designation as 'in' or 'out' will be the wrong one.

If these questions haven't yet begun to trouble you, fine. We could come to them later. But if, as I expect, they are already on your radar, I can only assume it will get worse, and I'd better address them right away.

*

Where is the theory most at risk? Perhaps the most serious challenge will come from an alternative theory—or class of theories—that starts off with an entirely different conception of what consciousness amounts to.

We've assumed all along that phenomenal consciousness is a *state of mind*. When you are conscious of seeing red, this means that you are entertaining *the idea* of phenomenal redness, evoked by red light arriving at your eye. But what if consciousness is actually a *state of matter*? What if the phenomenal redness is an intrinsic property of the brain activity associated with processing the visual information: a feeling that simply pops into awareness with no cognitive work being done to represent it and no work being done by it either?

It's a possibility I've raised already when discussing the neural correlates of consciousness, in Chapter 10. At that point, I summarily dismissed it as a bad idea. However, it's an idea that in one form or another keeps resurfacing.

Here are two distinguished scholars discussing consciousness, in 1971.[87]

Anthony Kenny (philosopher): 'It seems to me that on the view that Waddington has put forward there would be nothing nonsensical about the supposition that this mug should be conscious.' C. H. Waddington (theoretical biologist):

> I want to make a reply to some of the points just made by Kenny. Now, I am not a bit certain that the mug is not a little conscious. I said that I think you have to add to the definition of atoms something to do with consciousness, but I added that this consciousness is not going to be as highly evolved as ours...I am definitely not ruling out that there is some sort of thing allied to consciousness all through the world.

You might think that this way madness lies. Nonetheless, the notion that matter can somehow harbour phenomenal experience as an intrinsic property has a long history. For example, in the 1929 *Encyclopaedia Britannica*, under the entry for 'Consciousness', you could read about the 'Psychonic theory':

> One theory holds that each atom of the physical body possesses an inherent attribute of consciousness. If each atom, or, in later forms of this theory, each cell of the body emanates its own consciousness, then the 'self' must actually consist of an amalgam of all these tiny units of awareness... A second theory assumes that there exist, in the brain, special nerve cells capable of producing consciousness whenever activated... The psychonic theory, based on the correspondences between consciousness and inter-neuronic phenomena, suggests that consciousness occurs each time any unit of junctional tissue between individual neurones is energized. Units of junctional tissue are termed psychons, and each psychonic impulse is regarded as a single unit of physical consciousness. This theory is now under experimental investigation.[88]

In our own times, as we've seen already, similar ideas are being promoted by philosophers such as Philip Goff and Galen Strawson. And, more alarmingly, a sophisticated scientific version has been proposed by the neuroscientist Guilio Tononi.

Tononi's 'Integrated Information theory' (IIT) likewise begins with the 'correspondences between consciousness and inter-neuronic phenomena' and aspires to deduce the physical basis in the brain from the phenomenology of the experience:

> By starting from phenomenology and making a critical use of thought experiments, the IIT claims that: (i) the quantity of consciousness is the amount of integrated information generated by a complex of elements; (ii) the quality of consciousness

> is specified by the set of informational relationships generated among the elements of a complex.[89]

The theory proposes that consciousness is present to some degree wherever information is coordinated across the parts of a larger whole not only within living brains but also in any integrated system on any scale. Importantly, there doesn't have to be a defined subject *for whom* the experience is *about* what it's like. The experiencer is the system as a whole and the experience isn't about anything other than itself. IIT is described by the *New Scientist* magazine, in an article on 'Cosmic Consciousness', as 'our mathematically most mature theory of consciousness'. Christof Koch, who has helped develop the theory, has called it 'the only really promising fundamental theory of consciousness'.

I don't deny that the theory has a certain elegance. I'm not sure I fully understand the maths. However, I've no hesitation in saying that IIT fails to connect to the kind of consciousness under discussion in this book. Why should we engage with a theory of subjective phenomenal experience that unashamedly *leaves out the subject* and *leaves out the experience*?

The poet Coleridge wisely counselled that before you dismiss another person's beliefs it's as well to ask yourself *why* they have gone wrong. 'Until you understand a writer's ignorance, presume yourself ignorant of his understanding.'[90] But, in the case of Tononi and his followers, I'd say the source of their ignorance is clear enough. It's the mistake we identified earlier. Rather than asking how phenomenal properties are represented in the brain, they are seeking some feature of the brain that actually *has* these very properties. They are looking for the NC *of* C, not the NC *of representing* C.

True, it can sometimes happen that the vehicle for a representation does itself have the properties of the representandum, for example, the word RED written in red ink or the number FIVE represented by five dots. But there's no reason whatever to believe that this is true of how the brain represents sensations. As Dennett has put it, the phenomenal quality of a sensation of purple can be like 'a beautiful discussion of purple, just about a colour, without itself being coloured'.[91]

Earlier, I drew an analogy with the text that represents the story of *Moby Dick*. Suppose now that a literary theorist were to come up with 'Integrated Text Theory' to explain how the printed book can be the vehicle for the fictious story of the novel, starting from the premise that the printed book must have the same formal structure as the story it's about. 'ITT moves from phenomenology to mechanism by attempting to identify the essential properties of the story and, from there, the essential properties of the storytelling physical system.' It's not—I hope—a theory that would be taken seriously.

*

If panpsychism is out of contention, we needn't worry that we're talking about the wrong kind of consciousness. Nonetheless, we could still be talking about consciousness at the wrong scale. Suppose that consciousness of the kind we *are* talking about—sensations represented as having phenomenal properties—could exist in a simpler form and serve a lesser psychological function in lower animals. Not all the way down to coffee mugs, which aren't capable of representing anything at all, but all the way down to ants, say, which undoubtedly are, or even to bacteria, which may be. Then our theory, focusing on the existence of attractor states in the brain, might be setting the bar for sentience much too high.

The ipsundrum, as I've characterized it, could be the Rolls Royce for representing phenomenal experience. But perhaps less sophisticated vehicles, not involving attractors, could have evolved by a different route to become the basis for a different kind of phenomenal self.

Perhaps, but I can't see it. Let's not underestimate what our theory of attractors based on recursive feedback has going for it. (a) Such attractors can take on a vast range of properties with relatively minor adjustments to the circuitry, (b) they can be generated by the kind of feedback circuits that easily exist in the brain, and (c) there's a plausible evolutionary trajectory every step of the way.

But if, as I believe, this is the only viable theory on offer, this has a major implication for the distribution of sentience. It means that there must be a clear threshold between insentient and sentient animals. There won't be animals that are hovering half-way there.

Have you ever played with a 'Cartesian diver' (Figure 16.1)? As you release the pressure on the cork, the trapped bubble of air expands so the diver rises, the weight of water above is lessened, the bubble expands further, and the diver races to the top. When you increase the pressure, back it goes to the bottom again. Because of the positive feedback, there's no stable position in between. I don't know whether Descartes was truly the inventor of this toy. But, as a metaphor for why consciousness has to be either on or off, it might have appealed to him.

I actually see it as a strength of the theory that it rules out semi-sentient creatures. But I admit this puts me at odds not just with panpsychists but also with almost everyone else who writes about the distribution of sentience, including some I otherwise count as allies. I'll say some more to shore up my position and then offer a bit of a compromise.

The metaphor of the diver certainly won't appeal to Dennett. Twenty-five years ago, he went out of his way to reject the idea of a threshold:

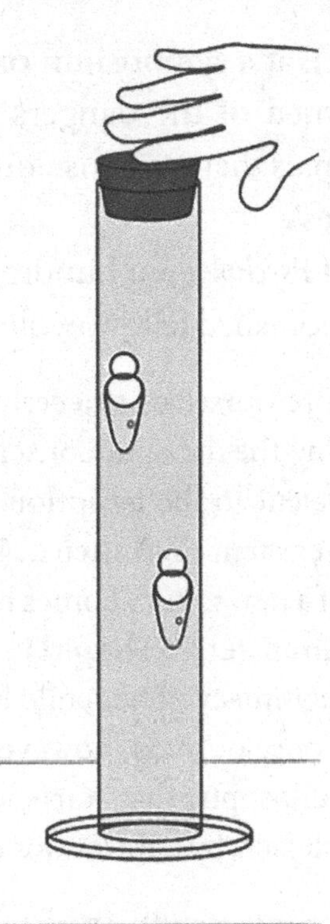

Figure 16.1 The Cartesian diver

> What does sentience amount to, above and beyond sensitivity? This is a question that is seldom asked and has never been properly answered. We shouldn't assume there's a good answer. We shouldn't assume, in other words, that it's a good question…Everyone agrees that sentience requires sensitivity plus some further as yet unidentified factor x…Here then is a conservative hypothesis about the problem of sentience: there is no such *extra* phenomenon. Sentience comes in every imaginable grade or intensity…the prospect that we will *discover* a threshold—a morally significant 'step', in what is otherwise a ramp—is extremely unlikely.[92]

He didn't at that point actually endorse this conservative hypothesis; he raised it as a possibility. But it's clear where he was leaning. In 2013, at a symposium on animal consciousness, he reportedly 'warned of the dangers of drawing sharp lines between some animals that are conscious like we are, and others that are just zombies'.

In his *Principles of Psychology*, a hundred years earlier, the great William James admonished fellow evolutionists:

> We ought therefore ourselves sincerely to try every possible mode of conceiving the dawn of consciousness so that it may not appear equivalent to the irruption into the universe of a new nature, non-existent until then... The fact is that discontinuity comes in if a new nature comes in at all. The quantity of the latter is quite immaterial. The girl in *Midshipman Easy* could not excuse the illegitimacy of her child by saying, 'it was a very small one'. And Consciousness, however small, is an illegitimate birth in any philosophy that starts without it, and yet professes to explain all facts by continuous evolution.[93]

I find the language interesting. Why is the idea of there being a discontinuous irruption of consciousness 'dangerous' or 'illegitimate'? Dennett's reference to a 'morally significant step' suggests what's troubling him. But whatever the reason for denying the possibility of a threshold, it can hardly be a scientific one. There's no reason to believe that in general *nature abhors discontinuities*. In *The Science of Logic*, the philosopher Hegel remarked:

> It is said that there are no sudden changes in nature, and the common view has it that when we speak of a growth or a destruction, we always imagine a gradual growth or disappearance. Yet we have seen cases in which the alteration of existence involves a transition, by a sudden leap, into a qualitatively different thing; an interruption of a gradual process, differing qualitatively from the preceding, the former state.[94]

The fact is that sudden breakthroughs—phase-shifts, catastrophes, tipping points, explosions—are a regular feature of natural phenomena and especially of the grandest ones: the Big Bang, or self-replicating life, or human language.

As I see it, the idea that phenomenal consciousness comes 'in every imaginable grade or intensity' is more a hope than a rational conclusion (though quite why it's something to hope for is not obvious to me). However, I reckon it has no support either in theory, in empirical data, or in human subjective experience. I've given my own reasons for believing there is indeed a 'factor x'—namely, the ipsundrum—that is either on or off. It seems to me quite certain that some animals are in effect 'zombies' with not a glimmer of phenomenal consciousness—and, in anticipation of later debate, I'll confess I'm glad of it.

This said, I'll now soften my position somewhat. My theory doesn't commit me to the view that all sentient animals are 'conscious like we are'. Nor, indeed, to the view that they are all *as* conscious as we are. I can agree with what seems to be the general opinion that sentient animals must differ in their range of sentience, so that, even if there is a step from insentience to sentience, *some sentient animals are more sentient than others.*

One obvious reason this may be so is that sentience is spread across different sensory modalities. This means that humans and other animals can be sentient several times over. To be visually sentient is one thing; to be tactually sentient is something more; to be auditorily sentient something more again. Therefore, sentience can add up across modalities.

Let's assume, as our theory would allow, that in the course of evolution sentience began in just one modality and was subsequently extended to others. Then, sentient animals would have become progressively more sentient over time, and there could indeed be a wide spectrum today. Suppose that a goldfish is phenomenally conscious of pain but not of any other modality and a frog is additionally phenomenally conscious of smell; then, the frog would be more sentient than the fish.[95]

Perhaps we can even draw meaningful distinctions between human beings. Helen Keller, being both blind and deaf, was arguably less phenomenally conscious than other normal humans. A patient who, as a result of brain damage, has lost visual sensation but retains the capacity for blindsight is less sentient than he himself was before.

Nonetheless, there remains a crucial step-like distinction between being sentient and insentient. For a formal analogy, following on from James, think of what it means to be a 'parent'. You cross the threshold to parenthood when you have a child. If you have more than one child, arguably you become more of a parent. Yet, in your own eyes and those of the world, you undergo an all-or-nothing change in status with the first child (even if it is a very small one). Psychologically—and possibly morally—I think the transition to sentience has been equally sudden and profound.

17

MAPPING THE LANDSCAPE

We couldn't open the discussion on where phenomenal consciousness exists in nature until we had a theory of how and why it might have evolved. The theory we've come up with may still be only a sketch. But I think we can be pretty sure it's a sketch of the right thing. We're not going to be blind-sided later by objections that we're talking about the wrong kind of consciousness or making unwarranted assumptions about its all-or-nothing status.

Let's proceed, then, with the two-step strategy for diagnosing sentience. First, decide who could not possibly be sentient on the basis of their brains and natural history; then, decide who most probably is sentient on the basis of specific tests.

The answer about who to exclude has been building steadily. The more we've deduced about the machinery required for phenomenal consciousness and the evolutionary history that has put it there, the more the field has shrunk. If we survey the animal kingdom, from earthworms to chimpanzees, it's increasingly clear that rather few animals meet the twin criteria of having brains that could deliver phenomenal consciousness and the kind of lifestyle that would make it advantageous. Therefore, if our arguments stand up, the default assumption must be that the vast majority of animals are insentient.

I realize that some people will find this assertion shockingly dismissive. 'You're saying the majority of animals are zombies!' If being a zombie means being an animal whose sensations lack phenomenal properties, yes, I'm saying that. But I should point out that this isn't such an insult as it sounds. For, within the theoretical framework we've established, the life of a zombie can still be quite a rich one. An animal whose sensations lack a phenomenal dimension can still be *cognitively conscious*: that's to say it can have introspective access to its mental states—perceptions, beliefs, desires, and so on—and show the intelligence that goes with this.[96]

What's more, an insentient animal can still have plain non-phenomenal sensations: basic brown-bag sensations, as it were. Let me summarize the history again. (a) First came *sentition*—an evaluative motor response to sensory stimulation. (b) Then came *sensation*, when animals discovered how to monitor this response so as to arrive at a mental representation of what the stimulation means to them. (c) Then, once the response was privatized and feedback loops established, came *phenomenal sensation*, with the representation taking on a whole new look.

How far different kinds of living animal have evolved along this path will have depended on what, if anything, they had to gain from it. There will be animals that have never got beyond stage one. Let's call these the 'Sensitives'. They respond to sensory stimulation but don't make a mental representation of it. I would expect this group to include animals with elementary uncentralized nervous systems, whose behaviour is largely reflex and does not involve creative processing of information: for example, sea anemones, starfish, earthworms, slugs.

Then, there will be those that have reached stage two and stayed there. We can call these the 'Sub-Sentients'. They do form mental representations of sensory stimulation and what it means but their sensations lack a phenomenal dimension. I would expect this group to include animals with developed brains that may be capable of intelligent behaviour that requires cognitive consciousness. They may be able to form quite complex societies. They will, however, have a limited sense of themselves as individuals and will not attribute selfhood or mental states to others: for example, honeybees, octopuses, goldfish, frogs.

Finally, there will be those that have reached stage three. These are the true 'Sentients'. They uniquely represent what's happening at their sense organs as having phenomenal depth. I would expect these to have large brains capable of supporting the complex sensorimotor feedback loops that create the ipsundrum. They will be highly intelligent, especially in the social sphere, and have a strong sense of their own individual selfhood: for example, dogs, chimpanzees, parrots, humans. And who else?

I'm going to narrow it down. I'm ready to argue that sentience is restricted to mammals and birds.

18

GETTING WARMER

What do mammals and birds have going for them that other creatures don't? What could be so special about their brains or their life-style that it allowed them to acquire sentience while others stayed behind?

We've seen how, in the history of sensation, the ground was prepared for making a leap to the phenomenal plane. Once sentition had been privatized, feedback loops were there for the taking—and sentience beckoned. However, it beckoned, as it were, from the upper terraces. The redesign of brain circuitry required to generate the ipsundrum and put it to use must have been quite a stretch. Could it have been achieved in short order simply by genetic mutation and recombination? I've wondered about it, and I confess I'm not so sure. In fact, I now want to make a rather different suggestion. I believe the breakthrough may have been occasioned by a dramatic shift at an environmental level: a change of circumstances that bumped the ancestors of today's sentients to the next stage, leaving the genes to catch up.

Birds and mammals have in common a physiological feature that distinguishes them from all other animals. They are *warm-blooded.* That's to say, they maintain a constant body temperature higher than the surroundings, typically 37°C for mammals and 40°C for birds.

I propose that warm-bloodedness played a double role in the evolution of sentience: on the one hand, it brought about changes in lifestyle that made sentience an essential psychological asset; on the other hand, it prepared the brain to deliver it.

*

A quick primer about warm-bloodedness. In both mammals and birds, it's a physiological state that is achieved by generating heat internally and having an insulating coat of fur or feathers to prevent heat loss. Fossil evidence shows that this capacity evolved independently in dinosaurs, the ancestors of birds, and cynodonts, the ancestors of mammals, at about the same time, 200 million years ago, during a period of major climatic upheavals.

Being warm-blooded is expensive. Maintaining a constant high temperature requires a big expenditure of energy. At 37°C, the human body is warmer than the mean annual temperature of any habitat on Earth. To keep this up, a human must eat nearly fifty times more frequently than a boa constrictor of equivalent size and consume up to thirty times more calories overall. Given such costs, there must have been big advantages or the trait would never have evolved.

In fact, the advantages are several. For one thing, as temperature goes up various bodily processes actually become more energetically efficient, so the costs can be partially offset. In particular, the cost of sending an impulse along a nerve decreases until it reaches a minimum at about 37°C. The result is that, although the overall running costs for the body go up with being warm-blooded, the costs for the brain are reduced. This means that mammals and birds can support larger and more complex brains with relatively little extra outlay of energy.

A separate advantage is that warm-bloodedness provides a defence against infections by fungi and bacteria. Cold-blooded animals such as insects, reptiles, and amphibians are plagued by fungal infections. But very few parasitic fungi can survive above 37°C. This means that mammals and birds are now largely free of them.

However, the fact that warm-bloodedness evolved when it did, at the same time in both classes of animal, when environmental temperatures were swinging wildly, suggests that the primary advantage was neither of these but rather the more obvious one that it allowed animals to ride out climatic changes and expand their geographic range.

Cold-blooded animals not only have to stay within relatively narrow geographic limits but also have their activity levels dictated from moment to moment by the ambient temperature. As the sun sets, or goes behind a cloud, the body of a cold-blooded animal such as a lizard chills and its muscles and nerves slow down; when body temperature drops too far, it becomes lethargic. By contrast, warm-blooded animals take their environment with them and so can be alert and active—feeding, socializing, travelling—both by day and night, winter and summer, high in the mountains or down on plains. The fossil record shows that at the time warm-bloodedness evolved many cold-blooded species, unable to cope with the fluctuating temperatures, became extinct.

As Claude Bernard put it, in his famous adage: '*La fixité du milieu intérieur est la condition de la vie libre.*' The constancy of the internal environment is the condition for a free life.

*

Now, what interests me is what a 'free life' means, not only for the *body* but also for the *mind*. As the bodies of warm-blooded animals became more autonomous, self-reliant, and self-contained, I imagine their *sense of self* did too. After millions of years in which their ancestors had their lives constrained by environmental temperature, they found themselves, as it were, let off the leash. In body and in mind, they were becoming increasingly autonomous agents, with the freedom to go where they would when they would.

I hear William James, celebrating the individuality of human minds: 'Absolute insulation, irreducible pluralism, is the law. It seems as if the elementary psychic fact were not thought or this thought or that thought, but *my* thought, every thought being owned.'[97] But insulation as a feature of the mind will very likely have begun with insulation as a feature of the body. Indeed, here's James again:

> Our entire feeling of spiritual activity, or what commonly passes by that name, is really a feeling of bodily activities whose exact nature is by most men overlooked... To have a self that I can care for, nature must first present me with some object interesting enough to make me instinctively wish to appropriate it for its own sake.[98]

A warm-blooded body is an object that must have been considerably more interesting to—and worth appropriating by—the self than a coldblooded one.

But this was just the half of it. I believe the change that warmbloodedness brought about in attitudes to the body and self was about to be amplified by what was happening at the level of brain physiology.

I've said little so far about what exactly might be required at the level of nerve cells to generate the attractors that are responsible for phenomenal consciousness. I won't pretend I'm ready to provide a detailed anatomical and neurophysiological model. Nonetheless, if I had to suggest an evolutionary change to the brain that would be conducive to establishing the feedback loops that create the ipsundrum, it would be (a) an *increase in the conduction speed* of nerve cells, effectively shortening the loops and putting motor and sensory areas of the brain closer in touch; coupled with (b) a *decrease in the refractory period* (the time-out) following a nerve cell firing, so that the cell can join in cyclical reactivation.

What a coincidence then, that an increase in the temperature of the brain would have been bound to have both these effects. It's a well-established fact of physiology that the functional characteristics of neurons change with temperature. It's been found for a range of animals—warm and cold-blooded—that the conduction speed for all classes of neurons increases by about 5 per cent per degree Centigrade while the refractory period decreases by roughly the same amount. This implies that when the ancestors of mammals and birds transitioned from a cold-blooded body temperature of, say, 15°C to a warm-blooded temperature of 37°C, the speed of their brain circuits would have more than doubled.[99]

*

We've remarked already on the 'lucky accidents' that have, at several points, played a part in the evolution of sensations. If warm-bloodedness played these key roles, first in changing the way animals thought about the autonomy of the self, second in preparing the brain for phenomenal consciousness, here was an accident as lucky as they come.

Cometh the hour, cometh the brain.[100]

19

TESTING, TESTING

We've been putting a lot of weight on arguments. Arguably, animals will have evolved to be sentient only if they need to have an individualistic sense of self. Arguably, phenomenal consciousness requires a large, warm-blooded brain. Arguably, it's only mammals and birds.

So, now we need the evidence. What empirical tests, if any, could settle the question of whether an animal actually is sentient? Is there some form of behaviour that, if the animal were to exhibit it, we could conclude for sure that the light was on inside? Or something else, that if the animal were to exhibit it, we could conclude the light was off?

Compare the question of whether a human possesses a capacity for colour vision. You're probably familiar with the Ishihara plates, used by opticians to test for colour-blindness in humans. With the 'Vanishing plate', only people with good colour vision can see the number twelve picked out in red dots among green ones of the same brightness. With the 'Hidden plate', only colour-blind people can see the number five, whereas for those with colour vision it's masked by the other coloured patterns they *can* see.

It would be asking too much to expect there to be diagnostic tests for sentience that give an equally clear-cut verdict. Sentience is not an accomplishment like colour vision, it's more a way of being, and a way of being that manifests itself in the first instance in private. The attribution of sentience is going to be a matter of inference rather than direct observation. Yet, while we shouldn't ask for too much, let's not ask for too little. In recent years, theorists who have given up on having a definitive test for sentience have been ready to settle for tests that, as I see it, are far too liberal and likely to lead to false positives.

The strategy that today is widely recommended by both philosophers and scientists is simply to test whether the candidates for sentience exhibit behavioural or psychological traits that are accompanied by phenomenal consciousness in humans. They tacitly assume that these behaviours are *caused* by phenomenal consciousness in humans and then go on to argue, by analogy, that what is true for humans is likely to be true for non-human animals also.

Philosopher Michael Tye, for example, appeals to a principle he attributes to Isaac Newton: 'The causes assigned to natural effects of the same kind must be, as far as possible, the same.'[101] So, if humans behave in a certain way in a certain situation because of their conscious experience of what it feels like, and if an animal in the same situation behaves in a similar way, we are entitled to assume that the animal's behaviour has a similar experiential cause. Primatologist Frans de Waal concurs: 'if certain capacities engage consciousness in humans, then they probably also do in other species'.[102]

The problem with this approach won't have escaped you. It's the assumption that when a particular example of human behaviour is accompanied by phenomenal consciousness, it must indeed be causally 'engaging' phenomenal consciousness and would not occur without it.

Tye considers pain behaviour to be a paradigm example. He observes that humans who are in pain typically show behaviours designed to counter the noxious event and summon help. 'It is a truism', he says, 'that we want to get rid of pain and that we behave in ways that reduce or eliminate it *because* of how it feels' (his italics).[103]

However, it's not that simple. As we discussed earlier, when, for example, you touch a hot stove, have a sensation of pain, and withdraw your finger, it's not clear that your being *conscious* of the pain—your knowing you have it—is playing any causal role at all. It's certainly true that you *have the impression* that your behaviour is caused by the pain sensation and, moreover, caused by the special quality of what it's like. You assume the behaviour wouldn't occur in the absence of the phenomenal feeling. That's the story you tell yourself. But, in reality, this story is most likely a user illusion—a way of making sense of things that doesn't truly reveal the causal structure of your mind.

Compare the case of free will. If you are asked to voluntarily move your finger, you'll have the impression that you do it by a conscious act of will. But this too is a story. It's been established experimentally by recording brain activity that your brain initiates the movement *before* you become conscious of willing it. If brain activity were to be recorded when you touch the stove, I've no doubt it would turn out to be the same: you initiate the withdrawal movement *before* you have the conscious sensation.

So, we should be sceptical about reasoning by analogy from what we know—or think we know—about humans. If it's open to question whether, in the case of humans, a particular behaviour is in fact being caused by phenomenal consciousness, we clearly shouldn't conclude that the same behaviour *can only be* caused by phenomenal consciousness in animals.

I'm not saying we should always be sceptical. Of course, I firmly believe that phenomenal consciousness, when present, can and does have causal effects—generating 'action-relevant changes in attitudes', as I put it earlier. There is an answer to 'and then what happens'. And often enough, in the case of humans, we do in fact read this causal relationship correctly. We recognize precisely what happens next and why.

When our story genuinely fits the facts about the cause of what happens next in humans, and we observe the same thing happening next in an animal, we'll be perfectly entitled to apply Newton's principle and conclude that the cause is the same. Indeed, I'd agree with Tye, de Waal, and others that, in this case, the argument from analogy is as good a way as any for inferring the existence of sentience in non-humans, possibly the best and only way. But, and it's a big but, we have first to get it right about humans.

We have, therefore, to home in on those changes in attitude, brought about by sensations, where we can be confident that it's the phenomenal properties of the experience that are responsible. And this means we may have to look beyond the 'typical' behaviours that immediately leap to mind. For instance, if it's pain we're interested in, we may want to consider paradoxical examples where people *welcome* pain because of its life-affirming properties, even as they want the pain as such to end.

I quoted Kundera, above: 'While it suffers, not even a cat can doubt its unique and uninterchangeable self.' Consider then, the case of 'Schrodinger's cat'. The cat is sitting in its box, suspended in an indeterminate state between life and death, until some-one opens the lid and takes a look. Suppose that, when the lid is opened, the cat *pinches itself to see if it is still alive.*

No doubt, this is more than we should expect of a cat. Nevertheless, I dare say it will have to be outliers like this, rather than mainstream behaviour, that will provide the clinching evi-dence for sentience.

*

So, to proceed, I'm going to bring up for discussion a number of behaviours that, if present, throw light on the existence of the *phenomenal self*: the 'I-thing', that in humans becomes an object of contemplation, attention, and ambition. By coming in at this level, we'll be looking for evidence of something we can be sure is an *effect*, not simply a corollary, of phenomenal consciousness. What's more, we'll be testing for the very thing that our theory suggests is what makes phenomenal consciousness an evolutionary success, so we'll be applying the tests that natural selection must have applied.

There will be three kinds of behaviour to consider: first, behaviours that are *enabled* by having a phenomenal self and *could* not occur without it; next, behaviours that are *promoted* by having a phenomenal self and *would* not occur without it; and then, behaviours that are *required* in order to maintain the phenomenal self and so would be *irrelevant* without it. The first two relate to what the phenomenal self *does* and the latter to what it *needs*. But evidence for any of them could confirm that the self *exists*.

Furthermore, there will be two levels of certainty about what we conclude from these behaviours. The behaviour we are looking at might be something that *only a sentient animal* could or would exhibit, so that this behaviour would be sufficient proof of sentience. Or, the behaviour might be something that *every sentient animal will be certain to exhibit*, so that failure to exhibit this behaviour would be sufficient proof of insentience.

I'll begin in the next chapter by considering behaviours that would seem to be essential to the *growth* and *maintenance* of the phenomenal self. Then, in the long chapter that follows, I'll turn to some of the *powers* the phenomenal self brings with it.

20

QUALIAPHILIA

In Chapter 14, when we discussed what it means for the phenomenal self to acquire its enduring identity, I likened the flow of sensations to a sequence of paintings hanging in a long picture gallery, stretching back into the past. As you grew up and the sequence unfolded, you will have come to recognize that these art works are all 'Yous'.

But now, let's take this right back to the beginning. In humans, the sensorimotor pathways of the brain don't develop their myelin sheaths until some months after birth, so the feedback loops that give sensations their phenomenal properties are probably not functional until then. This means that the first phenomenal sensations you encountered as a baby—the pains, smells, colours, and so on that were unaccountably *like* something—must have broken on your mind as a complete novelty. 'What's *this* about?' Presumably, the answer 'It's about me' will have taken some time to work out. You'll have needed practice to get the measure of being you.

Given the importance of the self to psychological development, we should expect young animals—young *sentient* animals—to be instinctively motivated to explore the territory of sensation in every way they can. To an outside observer, it will appear that the animal is gathering experience simply for the fun of it, indulging in exploratory play. But seen from within, every new experience will be adding a patch to the picture of what it's like to be *oneself*, helping to establish phenomenal identity.

*

So, which animals engage in sensory play?

1. Birds do. Young screech-owls pounce at leaves; young crows and jays pick up, inspect, and hide all kinds of shiny objects; young gulls and terns carry small items aloft and drop them, catch them in mid-air, and drop and catch them again. Hen chicks seek out new objects and poke or prod them. Parrots imitate an adult's behaviour, such as preening, as well as copying songs and sounds. Starlings tease or deliberately harass one another or taunt a domestic cat.
2. Mammals do. A similar list could be drawn up for human children, wolf cubs, kid goats, dolphin calves, or just about any young mammal.
3. Other animals don't. Reptiles, fishes, molluscs, insects, crustaceans, and others, play hardly at all. Nothing like this list could be drawn up for any of them. Perhaps this isn't so surprising in the case of animals who aren't notably intelligent. But we might think that octopuses, with their agile bodies and inventive minds, would be up there with dogs and parrots, pushing the limits of experience. The surprise is that young octopuses are much less playful than mice or sparrows.[104]

This seems counter-intuitive—until you take on board the argument that, if an animal shows no signs of going out of its way to grow and enrich its phenomenal self, it probably doesn't have a phenomenal self to grow and enrich.

This is a weak test for sentience. We can't say that if a young animal engages in exploratory play, it has to be sentient.

There are other reasons why play can help survival. Yet, we can certainly say that those animals that don't play are most unlikely to be sentient.

*

Let's move on from infancy and to a stronger kind of test.

The philosopher Tom Nagel has written:

> There are elements which, if added to one's experience, make life better; there are other elements which, if added to one's experience, make life worse. But what remains when these are set aside is not merely neutral: it is emphatically positive . . . The additional positive weight is supplied by experience itself, rather than by any of its contents.[105]

For Nagel, phenomenal consciousness as such is intrinsically valuable—valuable in a way that doesn't need justifying by reference to anything else. But, according to our theory, to value it is actually a biological imperative. Sentient animals have evolved to find experience instinctively valuable because those ancestors who strove to keep the phenomenal self afloat were the survivors.

So, the next test for sentience is simply this: does the candidate for sentience continue, as an adult, to nourish and affirm the self by seeking out the *kind of experience selves live by*? What kind of experience is that?

Nagel suggests that experiences are valued irrespective of whether they make life better or worse. We're in agreement on that. So, for example, while the sensations of sweetness and of sourness have very different implications for nutrition, both may be positive in supporting the phenomenal self. But it's hard to follow Nagel when he says that what matters is 'the experience itself' rather than 'any of its contents'. How can an experience be anything at all apart from its contents?

It would seem that what Nagel is getting at is none other than the difference between sensation and perception (although, like so many philosophers, he's not *au fait* with Reid's distinction). When you have a sensory experience, what gives it positive weight is the phenomenal character of the subjective sensation rather than any accompanying perception of the objective facts. This would be in line with our theory. But before we can develop this as a test of sentience, we need to say more about phenomenal content—why some experiences do more for the self than others.

*

We recognized in the earlier discussion that phenomenal consciousness adds up across sensory modalities: the more modalities involved in an experience, the more degrees of freedom, the weightier the experience. But it can also add up within a single modality: the more varied and the more aesthetically structured the stimulus field, the weightier again. There is more to being conscious when you see a scene in colour then when you see it in black and white; and when you look at a Kandinsky painting you are more conscious still.

Humans have an eye (and an ear, and a nose) for *sensory poetry*. That's to say, experiences where sensations play off against each other—rhyming, echoing, contrasting, chasing, complementing, thickening, calling attention to themselves, and throwing light on others. Whether any non-human animals delight in patterns of sensory stimulation in the way that humans do is something that's up for debate. But for humans, it's a strong and ingrained trait. We typically think of humans as an intellectual species, *Homo sapiens*, the cleverest species on Earth; but humans are also in several ways the most sensual species on Earth. From the time modern humans first appear in the archaeological record, around

100,000 years ago, their associated artefacts include flutes, wall paintings, sculpted figurines, body paint, decorative jewellery, collections of pretty shells. There's every reason to believe they were also creating art forms of which there's now no evidence: dances, cooked foods, flower garlands, sex.

Presumably, it didn't begin with artefacts. Our ancestors in the deep past must have taken their cue—just as we do today—from what was already out there in nature: the rhythms and harmonies of the natural world. Humans may be unusually sensual animals, but they are fortunate enough to live in a world that affords an astonishing range of things to be sensual about. Walk out the door and the poetry is there for the taking. Rainbows, clouds, mountains, sunsets, waves, snowflakes, stars, thunder and lightning, birdsong, ferns, blossoms, scents, butterflies, blackberries.

The poet Paul Valéry exclaimed: 'What speechless wonder that everything is, and that I am!'[106] Perhaps he might have said—on behalf of all sentient humans— 'because everything is, *so* I am'.

Could he have said that on behalf of non-human animals too? We'll see.

*

For humans, the outstanding example of the kind of experience that selves live by is music. The composer Michael Tippett asks:

> What does music really express? It's not about the sensations apprehended from the external world but about the intimations, intuitions, dreams, fantasies, emotions, the feelings within ourselves . . . Why we want this nobody knows, but human beings certainly do need it as part of something for which I think we must use the word 'soul'. We want our souls to be nourished, and unless they are nourished we are dead.[107]

This seems right. Except that music *is* about sensations first of all. If you doubt this, think about how little you would care for music if the same acoustical information were to arrive in a different sensory modality so that you were to perceive the same events but not feel them happening to you as sensations of sound.

True, it has sometimes been claimed that a musical score perceived visually can be the equal of the sound. A critic wrote of the fourth movement of Schubert's String Quintet, 'The very notes on the page look beautiful.' Some have even suggested that the sound gets in the way. The philosopher Adorno maintained that the ideal way to listen to music was silently, in the head; and he criticized Debussy for being too taken up with the way his compositions actually sounded—'fetishising of the material', Adorno called it.[108] The conductor Thomas Beecham witheringly remarked, 'The English may not like music, but they absolutely love the noise it makes.'[109]

But such rarefied opinions are worth quoting only to bring out the general rule: that with music you take pleasure in the *sensation of what's happening at your ears.* And you enjoy this subjective experience as it is, for its own sake: not because of what you *learn* from it, not because of what you have to *do* about it, but simply because it's good to *be there* as the subject of it. As Tippett says, it nourishes your soul.

Then let's return to the criteria I set out above for diagnosing sentience. Here, with listening to music, we have an example of a human behaviour for which, on the face of it, phenomenal consciousness is clearly causally responsible: absent the auditory qualia, most humans wouldn't care less. It's a behaviour that in many contexts has no obvious benefits beyond the feel-good effects on the self—and so not a behaviour for which as scientists we might want to provide a lower level explanation. In all, it's a behaviour that, if only we were to see it in non-humans, it would be churlish to deny that they too must be phenomenally conscious.

So are there, in fact, examples of non-human animals being attracted to music or anything equivalent?

*

There have been several experiments that have studied a related question: not so much whether animals like music, as whether they behave differently when music is being played. It's been consistently found that music of the right sort—slow and rhythmic—can have a calming effect. Kennelled dogs are less stressed in the presence of Reggae and Soft Rock music than Heavy Metal; their heart rate stabilizes and they are more likely to lie down. Cows give marginally more milk when exposed to slow classical music, like Beethoven's 'Pastoral Symphony'.

Such findings don't tell us whether the animals would choose to listen to the music if they had the chance. However, a few studies have gone part way towards this by looking at preferences for one kind of music as against another. In one study, the researchers set out to discover if domestic cats would show a liking for 'cat music' and prefer it to human music when they had a free choice of which of two speakers to approach.[110] Tests were conducted in the cats' homes, where the speakers were placed on the floor a few metres apart. The cats were allowed to roam freely and were watched by their owners to see if they responded differentially. The 'cat music' consisted of rhythmic and harmonious sound sequences specially composed so as to be within a cat's range of hearing and to contain cat-relevant content—purring and chirping. The human music for comparison was relatively simple classical music such as excerpts of a Fauré elegy and Bach's Ode on a

G string. The results showed that the cats were much more likely to orient to and approach the speaker playing the cat music.

It's promising. But does it show that cats *enjoy the sound* of cat music? It might do. However, with only the gross behavioural evidence to go on, it's hard to say. There are a variety of reasons why an animal might approach a source of sound. The cat might want to eat it, snuggle up to it, mate with it, or—most likely of all—simply find out more about it. Cat music could arouse a cat's curiosity more than human music simply because it sounds *more interesting* to a cat. But there's a big difference between interest in what's making a sound and pleasure in the sound as such. We humans don't listen to Bach's Ode because we want to find out more about it.[111]

The problem of interpretation is actually worse than this. As I came to realize with my own research on monkey aesthetics described in Chapter 4, evidence from preference tests—where an animal shows a tendency to stay longer with one stimulus than another—is not a reliable measure either of attraction or of subjective liking. As described above, I found that a monkey, given a choice between sitting in blue light or red, would spend three times as long in the blue. However, subsequent tests showed that this was not because it liked blue better but because it was slower to make the decision to try out the alternative. By the same token, the fact that a cat spends more time near the speaker playing cat music could be simply because it is slower to make the decision to move on.

In any case, as I now realize, all these studies have gone about things in the wrong way. Humans *seek out* musical experience: they don't simply respond to what's on offer, they go to some lengths to bring it on. Therefore, if we want to discover parallels with nonhuman animals, we must allow the animal to lead the way.

*

We want more naturalistic observational studies, where animals are left to their own devices and we watch to see what kinds of sensory experiences they go for. As a research strategy, this is bound to be more hit and miss. But when it hits, there's at least the possibility that we'll hit gold.

Fortunately, we have YouTube. Just as archaeology in Great Britain has benefited from the amateur metal-detectorists who have been doing the legwork pacing the ancient fields, so the study of animal minds has benefited from the army of video watchers who scan the internet looking for examples of animals showing human-like behaviour.

So let's join in. There are, indeed, a good number of videos on YouTube showing animals spontaneously approaching a human being playing a musical instrument. They come under titles such as:

Whales hang out to hear violin
Grazing cows rush to listen to accordion music
Karolina Protsenko is playing violin for a little squirrel
Man captivates horses with native flute.

Some of these cases are amazing. The squirrel, for example, approaches the young violinist, cocking its head, staring at the instrument. For us, watching the film, it's easy to imagine that the animal is approaching because it likes the sound, in the same way we might approach a busker in the subway. We can be sure that the animal is indeed taking the initiative. But the problem of interpretation remains. In every case, an alternative explanation is that the animal is *interested in*—in some cases obviously *puzzled by*—what's going on. Suppose the tables were turned and you were to come across a kangaroo playing the trombone. No doubt you'd go and have a closer look but not for the sake of the musical sensations.

When could we be sure it's more than curiosity? If the animal has been there before and already knows quite well what to expect, then we'd be on stronger grounds for arguing that it's pleasure rather than interest that's at work. With humans listening to music, this is often the case. Your pleasure in a Beethoven sonata won't be diminished by the fact you've heard it before (it might, indeed, be increased). However, so far as I know, there's no evidence of an animal coming back to listen to the same music again and again.

In other contexts, animals do, of course, return for more of sensations they are fully expecting. Honey gathering by chimpanzees provides a dramatic example. Chimps in the Congo climb high into the forest canopy to attack bees' nests with clubs. In the words of one of the researchers, 'The nutritional returns don't seem to be that great. But their excitement when they've succeeded is incredible, you can see how much they are enjoying tasting the honey.'[112] The chimps are clearly achieving a sensory high. They aren't doing it to learn anything. The combination of sweet tastes, painful bee stings, muscular effort, banging clubs, shrieks of excitement, and the ongoing danger creates a symphony of sensation.

Yet, the trouble is that here there clearly is a lower-level explanation. Even if the nutritional returns aren't great, they aren't negligible. All we need assume is that the chimps are programmed to seek sugary sensations when they are short of calories, and in this case it's gone over the top. What we really want are examples more like music, where we can be confident the sensations are bringing their own reward and not being sought as a proxy for something else.

Chimpanzees' behaviour at waterfalls would seem to come closer to this. At the Gombe Stream reserve in Tanzania, adult males have been observed to visit a waterfall in spate, as if deliberately seeking nothing other than a consciousness-expanding experience. Bill Wallauer, the National Geographic cameraman, describes one occasion:

> Freud [the alpha male] began his display with typical rhythmic and deliberate swaying and swinging on vines. For minutes he swung over and across the eight-foot to twelve-foot falls. At one point, Freud stood at the top of the falls dipping his hand into the stream and rolling rocks one at a time down the face of the waterfall. Finally, he displayed (slowly, on vines) down the falls and settled on a rock about thirty feet downstream. He relaxed, then turned to the falls and stared at it for many minutes. It was one of those times that I would give body parts to know what was going through a chimp's mind.[113]

Jane Goodall certainly sees it as an awesome experience for the chimpanzee: 'Why wouldn't chimps have feelings of some kind of spirituality…which is really being amazed at things outside yourself?'[114]

Or, as we might put it, being amazed *at* yourself, revelling in your own existence.

*

Do animals other than chimpanzees get inspired by natural phenomena like this? The questions are out there on the internet. The answers—to take seriously—are much fewer. There's no evidence that any animals go out of their way to enjoy sunsets, rainbows, cloud formations, or drifts of bluebells. Nor, for that matter, do they climb mountains just because they're there. But, on their own level, they certainly exploit the opportunities nature affords for sensory adventure.

On YouTube, you can see some wonderful examples. Swans surf the incoming waves on a beach and go back for more.

A rook jumps onto a tin plate and sledges down the snow-covered roof of a house, takes the plate back up to the top, and does it again. Dolphins ride the bow wave of a ship. Monkeys jump from a high perch into a pool with a huge splash. Galahs—small Australian parrots—fly into a whirlwind, get hurled upwards, screeching loudly, emerge at the top, and fly down to the base to re-enter for another go. Buffalo run onto and slide across an icy pond, excitedly bellowing as they do so. A dog drags a toboggan to the top of a hill, jumps on to it, slides a hundred metres down a snowy slope, turns around, and has another go.[115]

It's tempting to see a common theme here. These activities—across species and environments—all involve becoming *ungrounded*: defying gravity, achieving weightlessness. We could put humans in there too. Humans love to ski, to glide, to dive, to swing, to ride the roller-coaster. It's as if there's an archetypal urge to leave their bodily mass behind. What's more, humans regularly *dream* of flying.

Dreams in which you overcome gravity and float above the ground or even fly like a bird are remarkably common. They have been recorded from people of all cultures and from far back in history. Sigmund Freud suggested that flying dreams are shaped by childhood memories of playing on a swing or being swung round by the arms.

Anthropologists have remarked how often people regard their dreams as evidence of the mind's independence of the material body. In a study where people testified to what the flying dreams mean to them, Claire Mitchell asked participants whether 'they were themselves' during the dream.[116] All affirmed that they were. However, as she says, 'I realised they were actually describing being "more than themselves".' In the words of one participant, 'I am me. I am absolutely me, but I am me without the weight of the world.' And in those of another, 'I don't feel mortal.' Do non-human animals—and not just birds!—dream of flying? I see no reason to deny it—nor that it could have positive effects on their self image.[117]

*

I've saved for last a particular kind of sensation seeking because it seems to be categorically different. It's where an animal, rather than looking for sensory stimulation from outside, creates the sought-for stimulus internally by manipulating its own body. I'm referring to self-pleasuring, masturbation.

Masturbation is a widespread practice in both mammals and birds. In terms of frequency, humans lead the field: young men masturbate to climax on average every three days and women are not far behind. Bonobos run humans a close second. But donkeys do it, penguins do it, bats do it. In fact, just about everybody does it. Perhaps the most creative form of masturbation is that of the male bottlenose dolphin, which has been observed to hook up an eel from the sand with its penis and proceed to float on its back with the wriggling eel doing the business.[118] Almost as inventive are monkeys and chimpanzees that use live toads as sex toys. Erica Jong notoriously extolled the zipless fuck: 'The zipless fuck is absolutely pure. It is free of ulterior motives . . . No one is trying to prove anything or get anything out of anyone. The zipless fuck is the purest thing there is. And it is rarer than the unicorn. And I have never had one.'[119] But, arguably, the self-fuck gets close to this ideal. As Quentin Crisp quipped in a film for the BBC (though the line was subsequently censored): 'Sexual intercourse is a poor substitute for masturbation.'

I remarked in Chapter 14 that orgasm, for humans, brings the phenomenal self sharply into focus. And, though the experience revolves around bodily sensations, it can have a sublime unbodied quality—just as in those dreams of flying, but even more so. Would it be too much to suggest that the same could be true for non-human animals?

It's true that, as with honey gathering, there could be a lower-level explanation of the benefits. Masturbation does, indeed, mimic sexual intercourse. So, even a non-sentient animal might very well be programmed to do it reflexly when the opportunity presents itself. We shouldn't be surprised if there are insentient animals that sometimes masturbate—having mistaken it for intercourse. The behaviour has, in fact, been observed in tortoises, lizards, and frogs. I don't know whether a brainless frog would do it, but I wouldn't be surprised.

However, there are several reasons to believe that masturbation in mammals and birds can't be simply an overspill from reproductive sex. Humans regularly start masturbating as small children and continue into old age. For many animals, as well as humans, orgasm is more commonly achieved by masturbation than by intercourse. Masturbation is often a solitary activity, like listening to music through ear-phones. The sensory pleasures—at least for humans—can be the equal if not greater than those from partnered sex.

From an evolutionary perspective, this would all be very odd unless there were non-sexual benefits. There's no question that natural selection first made orgasm pleasurable so as to promote sexual intercourse, leading to the birth of offspring. But why go on to design orgasm to have such exotic qualities? Why make self-pleasuring so readily available and attractive when it can be so distracting from other more worthwhile activities? Let's at least entertain the possibility that orgasm has long played a part in the birth and sustenance of the self. And that the properties of orgasm have been tweaked in that direction.

As we saw earlier, the sensation of orgasm is in a class of its own. No other sensation remains so closely linked to a motor response. The 'coming' is a real muscular event. But, as with other sensations, the experience arises from reading the motor command signals, not the response as such. The sensation can still be present in someone whose spinal cord has been cut, so that ejaculation, uterine contractions, and so on are not actually occurring. Several aspects of the responses in both men and women—for example, the waves that seem to spread beyond the genitals—have nothing to do with conception. Have they been introduced into the design of orgasm specifically to add complexity and harmony to the experience? Not to put too fine a point on it, is orgasm self-made body music?

*

We've argued that the phenomenal self is an asset that needs nurturing. Sentient animals require regular doses of phenomenal experience to keep their selves afloat. When animals seek sensations for sensations' sake, it's strong evidence that they are in fact sentient. Waterfall displays by chimpanzees, tobogganing by dogs, masturbation by ducks . . . all point this way. It's hard to imagine what these animals could be seeking if it's not phenomenal experience.

This is a strong test for sentience. But it's a test that sets a high bar, and we should be aware that not all sentient animals are going to pass it. We've looked for evidence that animals seek experiences that have no practical value in order not to count cases where the behaviour might be motivated by, say, simple curiosity or hunger for calories or sex. This means that we haven't counted cases where the behaviour could be rewarding on several levels at once: both as food for the self and food for the body. The chimp quest for honey could be such a case, as could orgasm in the course of reproductive sex.

But we might well imagine that there are sentient animals who actually get all the experience they need in the course of everyday pursuits: eating, or grooming, or smelling each other's bottoms. In that case, these animals might never meet our criterion of seeking sensation for sensation's sake. They won't *have to* masturbate or go sliding on the ice. If we find they do, well and good, we can say they are sentient. But if we find they don't, we can't say they are not.

This goes for humans too. Humans don't *have to* listen to music. Tippett may be right that music nourishes the soul, and unless humans' souls are nourished they are dead. But the souls of those humans who don't listen to music are presumably not dead, just a bit less alive.

*

So, let's make an interim assessment of the evidence under the heading 'Qualiaphilia'.

Mammals and birds universally play as infants. Some (though it may not be all) go on to seek sensory experience for its own sake. Animals other than mammals and birds play very little by comparison. And none, so far as I know, seek out sensory highs.

This is what we'd have predicted. Play is something we would expect to see in all animals that are indeed sentient. So the absence of play tells against sentience (even if its presence doesn't confirm it). Seeking sensation for sensation's sake is something we would expect to see *only* in animals that are sentient. So, its presence confirms sentience (even though its absence doesn't disprove it).

But so far we've looked only for evidence of behaviours that are related to growing and sustaining a phenomenal self. We've yet to discuss behaviours that would provide positive proof that animals are putting this self to use in the conduct of their lives.

21

THE SELF IN ACTION

At several points already, we've discussed how it works out for humans: how having a self that is built around the properties of bodily sensations shapes a person's psychology, adding to the sense of their own worth and changing their attitudes to others.

For humans, the phenomenal self becomes the enduring entity that constitutes your 'I'. It's this 'I' that you wake up to each morning and put to bed at night; this 'I' that owns your beliefs, desires, and actions and gives them a narrative coherence; this 'I' that figures in your memories and dreams; this 'I' that carries forward your hopes, fears, and ambitions; this 'I' that is the template for imagining the selves of other people.

So, how will this play out in non-human animals? What evidence is there that animals have anything like an 'I', understand its structure, attribute such a self to others?

These are the right questions—and potentially good tests of sentience. But can we even ask them about animals? To show how these questions can be pitched at an appropriate level, I'll take them to be questions about a particular animal that I know well, who can then become a reference point for other species.

*

My dog, Bernie, is a four-year-old standard poodle. Everything suggests he is as sentient as they come. I say this on the evidence of his irrepressible *joie de vivre*. True, I haven't seen him riding a toboggan. But when I so much as hint that he is about to be taken for a walk, he springs a metre high in joyous anticipation. Once out on the path, he stretches his head to sniff the air, then races off in a madcap pursuit of anything that moves—or of nothing at all. He finds a stick and brings it to me to throw for him, again and again. When we come to the river, he leaps in. He grabs a mouthful of weed and tosses it the air, then scrambles back up the bank and spins in circles like a whirling dervish. Once we get home, he chases the cat into a tree just for the hell of it, then follows me into my study, where he takes up a post on the window seat to watch the world go by. Every so often, he saunters over, puts a paw on my chair, and asks to be stroked. He hears the postman's knock, sprints to the front door, and barks like mad.

Qualiaphilia, thy name is Bernie. But let's ask how he—and other likely candidates for sentience—would fare on some crucial functional tests.

Figure 21.1 Bernie

Does Bernie have an idea of his own continuing identity—anything equivalent to an 'I' that he carries forward?

He certainly knows his own name when it's spoken. If he could express himself in words, I'm sure the word 'Bernie' would be one of them. A dog called Stella, in America, has been trained to press one of twenty-nine buttons on a keyboard with its paw, each producing a different spoken word. One of the buttons makes the sound 'Stella'. She is said to press sequences such as 'Stella Walk Outside' and then run to the door.[120]

Animals of several other species have been taught to communicate symbolically with humans: Washoe the ASL-signing chimpanzee, Alex the talking parrot, Koko the signing gorilla, Sarah the lexigram-writing chimpanzee. Although none have ever reached human-like levels of language, all have quickly learned to make use of a sign or symbol to refer to themselves. And all have used this self-symbol to express first-person desires and emotions.

I had the opportunity in 2001 to observe an African grey parrot, called N'kisi, who lived with his owner Aimée Morgana in Harlem, New York. Aimée had raised him in her home since he was a chick, coaching him intensively to interact through spoken English. At three years old, he had a vocabulary of several hundred words that he spontaneously used to refer to things and people and sometimes strung into meaningful sequences.[121]

On the day I visited, N'kisi was flying freely around the sitting room. There was another young parrot, Endora, in a cage in the corner. Aimée had told N'kisi to keep away from Endora. N'kisi sat on a shelf looking longingly at her. I heard him say out loud, apparently as a self-instruction, 'N'kisi, No. N'kisi, No.'

But eventually he couldn't help himself and he flew down and landed on the cage. Endora attacked his foot and bit one of his toes, causing a nasty cut. N'kisi retreated to the shelf and pityingly held up his injured foot. He looked at Aimée. 'N'kisi hurt', he said, 'Want to fly over.' 'OK. Poor N'kisi', Aimée said, and he flew to land on her shoulder.

If Bernie hurts his foot, he too will come over to me to have it tended to. Suppose Bernie could express the idea 'Bernie, hurt; want to get petted.' The question is: what would the term 'Bernie' refer to? Presumably, he would have to be thinking of something that is the subject of the hurting and the wanting. Indeed, he'd have to be thinking of his subjective self. 'I' hurt, 'I' want.

What other evidence is there that Bernie has a concept of 'I'? Research that directly addresses the self-concept of dogs is limited and not very revealing. Dogs fail the mirror-mark test, which tests whether they can learn, with practice, that the face of the dog they see in a mirror—which now has an unexpected mark on it—corresponds to their own face. They pass a kind of mirror smell test, which tests whether they realize by sniffing that a patch of urine is their own urine and therefore relatively uninteresting compared to another dog's urine.

But these tests have little if any bearing on the existence of the dog's subjective self. The mirror-mark test, although often cited as evidence of 'conscious selfhood' in animals, is more a test of intelligence than of selfhood. It requires that the animal is able to learn with practice that the face in the mirror is an *image* of its own face (which, of course, it has never seen directly).

Chimpanzees, elephants, and dolphins are among the few species of animal that can do this. But, even if an animal *can* do it, it's not clear what having a face has to do with having a subjective self. As Milan Kundera has written, 'the face, that accidental and unrepeatable combination of features...reflects neither character nor soul, nor what we call the self. The face is only the serial number of a specimen.'[122] Bernie, as it happens, has a microchip implanted under his skin. If the number were to be displayed on a screen, I expect he could be trained to recognize it as *his* number. The number could effectively become a visual version of his name, along with the acoustical version 'Bernie'. But his self wouldn't reside in the number any more than it does in the sound or in the face.

The self doesn't reside in the word 'I' either. The word 'I' too is a name. Boringly, it can be the name of the speaker's body: 'I am six feet tall', 'I am going shopping'. But interestingly, and usually, it is in fact the name for the speaker's subjective self. 'I feel', 'I think', 'I remember'. If Bernie could indeed learn to use his name to communicate his states of mind, as Stella apparently can, that would be a much better test of conscious selfhood than any mirror test. I say 'if' he could. But when he comes over to me and lifts up his paw, I believe he is in fact doing just that. He is naming himself in the act of approaching me. His action is speaking as loud as any words.[123]

Does Bernie time travel with his 'I'?

After N'kisi had been bitten, I heard from Aimée that in the days following he repeatedly referred back to the incident, saying 'She cut my toe.' This would seem to be a clear case of his reliving an episode as he experienced it. He had travelled, as N'kisi, back in time. In fact, there's evidence of episodic memory that's still more remarkable. At one point, Aimée took him with her when she went to stay with relatives in Chicago. While they were there, to everyone's disappointment, he hardly said a word. Once he got home, however, he talked about one thing after another that had happened on the trip.

Does Bernie travel back like this? The evidence is more prosaic. But here's a small example. I usually give him his food before we go for a walk. If, after I've filled his bowl, I stand at the door waiting to go out, this puts him in a quandary: finish eating or set out straight away? Usually, he decides to wolf down the food; but sometimes he's so keen on the walk that he leaves with me before he's emptied the bowl. What I've observed is that, when we return home half an hour later, he clearly remembers how he left things: if he hadn't finished the food, he races back to the bowl; if he had, he doesn't bother.

If Bernie can remember facts like this, whether the bowl was empty last time he was there, can he remember his own actions—what *he did* last? The ability of dogs to remember their own actions has been put to the test experimentally.[124] Ten dogs took part in the study. Initially, each dog was taught by its owner to perform several actions on command, and then, if given the command 'Repeat!', to do it over again. So, in the test session, the owner waited until the dog spontaneously performed some new action—such as picking up a toy, drinking from a bowl, or jumping on a sofa; the owner then called the dog over and, after a delay, unexpectedly gave the command 'Repeat!' When the delay was twenty seconds, the dogs successfully repeated the preceding action on 70 per cent of trials; when it was one minute, the dogs repeated the preceding action on 65 per cent of trails. When the delay was one hour, during which the dog sat in a cage, they were still successful on 35 per cent of trials. The authors of this research are careful to say that they've found evidence only of 'episodic-like memory' and not to claim that the dog's conscious self was necessarily involved. But I'm ready to say it for them on behalf of Bernie: it's reasonable to conclude that he does indeed keep a running record of his recent history, to which he can return in the first-person as 'I'.

Now, if an animal can project its self backward in time, remembering how things *were*, it's probably not a big step up in brain power to be able to go forward and imagine how things *will* be. I don't know of evidence that dogs can do this. But there's good evidence for forward planning by other species. Research from Nicky Clayton's lab in Cambridge has shown that birds of the crow family can learn what the future holds for them and will make preparations in advance to meet their anticipated needs. Scrub jays will, for example, store food in a location where they expect to find themselves hungry the next day. In like fashion, chimpanzees can learn what tools they will require to solve a problem and will bring them along. There was a chimpanzee in the Malmo Zoo who took a particular delight in throwing rocks at visitors on the other side of a moat. At night, when the zoo was closed, he would replenish his pile of rocks so as to be ready for the next day's encounter.

So, if you can project your self backwards and forwards in time, how about sideways? If you can imagine being in those shoes of yours in the future, perhaps you can imagine being in someone else's shoes right now. As William Hazlitt observed two hundred years ago, 'The imagination by means of which alone I can anticipate future objects must carry me out of myself into the feelings of others by one and the same process by which I am thrown forward as it were into my future being.'[125]

He makes it sound easy. However, in reality it may not be so. Even if you have a direct line to your own 'I' as it was in the past and will be in the future, it doesn't follow that you will be able to make the leap into the 'I' of someone else. Is this something of which Bernie is capable?

Does Bernie regard others as having an 'I' in their own right?

He certainly thinks of the humans, dogs, and cats that constitute his social world as individuals. He treats them differently and has different expectations of them. He even knows a good few by their proper names. But, of course, this need not mean he regards them as having an 'I' *like his*. Does he understand that beings other than himself can be the subject of experiences that are as real and personal for them as his are for him?

We've noted that the word 'I' typically names the speaker's self. As such, it's a *personal* pronoun, one of those words like 'you', 'me', and 'he' whose referent changes according to who is speaking it. When you hear another person speak the word 'I', you tacitly assume that they are pointing to a centre of self belonging to them, not you. You might think this would be too difficult an idea for any non-human animal to come to terms with. However, in a clever experiment, researchers in Kyoto did in fact successfully teach symbols for personal pronouns to Ai, a female chimpanzee.[126]

Ai and a human companion sat together at an interactive computer console and played a match-to-sample game. Each round started with one of them pressing a button to bring up a symbol on the screen that stood either for ME or for YOU. Two seconds later, this was followed by two photos, side by side of Ai and of her human companion. If it was Ai's turn to start and ME came up, she had to touch the photo of Ai, while if YOU came up, she had to touch the photo of the companion. If it was the companion's turn to start and ME came up, Ai had to touch the photo of the companion, while if YOU came up, she had to touch the photo of Ai. Ai mastered this task within a few daily sessions.

And she continued to get it right when the human companion was replaced by one of nine other people.

It's remarkable evidence of the chimpanzee's ability to understand the *conditional* meaning of a symbol. 'If I bring up the symbol ME it stands for the photo of me, but if you bring up the same symbol ME it stands for the photo of you.' Bernie is not as clever as a chimp, and I'd be surprised if he could learn to do this with abstract symbols. However, suppose it were to be a question of understanding the conditional meaning of a *bodily expression*: 'If I say Ouch!, it means I'm in pain, but if you say Ouch!, it means you're in pain.' This could be a task for which he is much better prepared.

In fact, in this case, I've no doubt he would get it. As I said, if he's hurt and whimpers, he expects me to comfort him. But if *I* whimper, he comes and licks *me*. This would seem to imply that Bernie understands that someone other than himself can be the subject of personal experiences that are as significant for them as his are for him. But does it? If so, how far can he discern what those experiences are?

Is Bernie a natural psychologist? Can he read minds?

The answer, I think, is 'Yes, up to a point.' But I don't think it can be an unqualified 'Yes', for Bernie or any other non-human animal. To explain my caution about this, which may seem to go back on what I've said earlier, I need to say a bit more about the history of theories about 'theory of mind'.

I described, in Chapter 9, the background to my ideas about 'natural psychology'. My observation of mountain gorillas had set me thinking about social intelligence and the kinds of demands it makes on an animal's ability to reason about behaviour and relationships. How can an animal possibly manage to be—as it must be—a psychologist? What kind of model does it have of how another creature's mind works?

One answer—it would have been the conventional one fifty years ago—would be that animals go about the problem in the way that behaviourist scientists have done by painstakingly assembling observational evidence and trying to make sense of it. But that, I realized, would leave them struggling: too slow and too difficult. There must be some other way. And, of course, we humans know what the other way is. What *we* do is to take a short-cut: we use our own minds as a model. That's to say, we make use of our introspective knowledge of what it's like to be ourselves in order to imagine what it would be like to be another person.

Humans, by virtue of their phenomenal consciousness, are able to be *mentalists*, not *behaviourists*.

But if this is how humans do it, we might guess that there are animals who are capable of doing it this way too. Although my immediate concern was with gorillas, I soon wanted to include many other highly social species: rats, and whales, and, of course, dogs (I confess I didn't at that point include birds).

I set out these ideas in a lecture for the British Association for the Advancement of Science in 1977.

> The trick which Nature came up with was introspection; it proved possible for an individual to develop a model of the behaviour of others by reasoning by analogy from his own case, the facts of his own case being revealed to him through examination of the contents of consciousness... [A social animal could never succeed as a behaviourist.]... If a rat's knowledge of the behaviour of other rats were to be limited to everything which behaviourists have discovered about rats to

> date, the rat would show so little understanding of its fellows that it would bungle disastrously every social interaction it engaged in... Behaviourism as a philosophy for the *natural* science of psychology could not, and presumably does not, fit the bill.[127]

Then, in 1978, David Premack and Guy Woodruff reported in the journal *Science* a study with their chimpanzee Sarah that appeared to be a brilliant test of this ability to attribute mental states.[128] Sarah was shown videos of a man who was having difficulty in solving a problem: for example, he was unable to reach a bunch of bananas, unable to open the door to the cage, unable to get water from a hose. She was then shown two photos, one of which depicted a solution to the specific problem: a key to the lock, some boxes to stand on, a water-tap to turn, and so on. With seven out of eight problems, Sarah chose the correct solution, suggesting that she could indeed see things from the man's point of view.

Their paper was entitled 'Chimpanzee Problem-Solving: A Test for Comprehension' and, as they explained, 'This test... permits the animal's knowledge about problem-solving—its ability to infer the nature of problems and so recognize potential solutions to them—to be examined.' Yet, surprisingly, the authors did not at first make much of the fact that this was specifically a test of *mental* comprehension.

I had been in correspondence with Premack the previous year, after writing a review of his book, *Intelligence in Apes and Man*, for *Science*. I sent him a copy of that lecture, entitled 'Nature's Psychologists'. He didn't pass comment. But it soon became clear that we were thinking on parallel lines. Indeed, within the year, he and Woodruff came out with a new paper in which they reported the same experiments with Sarah under a different title: 'Does the Chimpanzee Have a Theory of Mind?'[129] The emphasis now was all on mind-reading: 'We are less interested in the ape as a physicist than as a psychologist... In this paper we speculate about the possibility that the chimpanzee... imputes mental states to himself and to others.' And they concluded, 'The ape could only be a mentalist. Unless we are badly mistaken, he is not intelligent enough to be a behaviorist.'

This was the first use of the provocative phrase 'theory of mind'. The paper, with its elegant experiments, sparked a wave of further research into the mind-reading abilities of other animals and indeed humans. Behavioural scientists hoped to be able not only to confirm the results with Sarah but also to extend them. But, dismayingly, it didn't happen. In fact, subsequent attempts to show that chimps or any other non-human animals can accurately discern the thoughts and feelings of others have been frustratingly inconclusive. Writing in 2015, Celia Heyes concluded:

> Research on mindreading in animals has the potential to address fundamental questions about the nature and origins of the human capacity to ascribe mental states, but it is a research programme that seems to be in trouble. Between 1978 and 2000 several groups used a range of methods, some with considerable promise, to ask whether animals can understand a variety of mental states. Since that time, many enthusiasts have become sceptics, empirical methods have become more limited, and it is no longer clear what research on animal mindreading is trying to find.[130]

*

What went wrong? I think the problems began with the word 'theory' in 'theory of mind'. It raised false expectations about the sophistication of attributing mental states. In my own writings, I had proposed something much simpler. As a natural psychologist, you read another's mind when you assume they are having similar thoughts and feelings to those you had when you were in the other's situation. You are reminded of what it was like for you, and this gives you a sufficient basis for predicting what comes next. No theory needed. But, in Premack and Woodruff's formulation, mind-reading became something much grander. You work out what it's like to be the other by rationally thinking it through. You work out, say, that the man is trying to escape and can't find the key, or that he mistakenly believes there's a snake under the bed, or that he can't see the banana from his position in the room.

This emphasis on reasoning was compounded when Premack and others harked back to Dennett's notion of the 'intentional stance' that he introduced in 1971 as a way of characterizing how humans go about predicting the behaviour of other 'intentional systems':

> Here is how it works: first you decide to treat the object whose behaviour is to be predicted as a rational agent; then you figure out what beliefs that agent ought to have, given its place in the world and its purpose. Then you figure out what desires it ought to have, on the same considerations, and finally you predict that this rational agent will act to further its goals in the light of its beliefs. A little practical reasoning from the chosen set of beliefs and desires will in most instances yield a decision about what the agent ought to do; that is what you predict the agent will do.[131]

Now, it's certainly possible that there are non-human animals that have the brain power to do this: to 'figure out' rationally what's going on in another's mind. But the rather surprising fact is that, when they've been put to the test, none—not even chimpanzees—seem to be much good at it.

I hasten to say that even if animals don't have a theory of mind in Premack's sense or adopt the intentional stance in Dennett's, they may still be capable of being natural psychologists of a less theory-based kind. As we saw earlier, there are less demanding ways in which you can exploit your own experience to gain insight into others. Even if you aren't able to put yourself in the other's place so as to work out what they are thinking or feeling right now, you can still use yourself as a model to get a generic understanding of what's possible for them. In particular, if you can only understand the scope of the other animal's sensory capacities—that they see, or, hear, or smell, or taste, or touch—in the same way you do, you'll be able to get a good fix on the kind of actor that you're dealing with.

You'll recall that I've raised this issue more than once already, when discussing the advantages of having a phenomenal self and when musing about how Helen, the monkey with blindsight, might have had trouble understanding what it's like for another animal to see. I developed the case of Helen as a thought experiment in 'Nature's Psychologists'. It's now especially pertinent, and I'll quote at greater length.

> [Let's consider] the hypothetical case of a monkey who has been operated on soon after birth and who therefore has never in his life been conscious of visual sensations. Such a monkey, relying only on perception, would presumably develop the basic capacity to use visual information in much the same way as does any monkey with an intact brain; he would become competent in using his eyes to judge depth, position, shape, to recognize objects, to find his way around. Indeed, if this monkey were to be observed in social isolation from other monkeys, he might not appear to be in any way defective. But ordinary monkeys do not live in social isolation. They interact continuously with other monkeys and their lives are largely

> ruled by the predictions they make of how these other monkeys will behave. Now, if a monkey is going to predict the behaviour of another, one of the least things he must realise is that the other monkey himself makes use of visual information—that the other monkey too can see. And here is the respect in which the monkey whose visual cortex was removed at birth would, I suspect, prove gravely defective. Being blind to the sensations of sight, he would be blind to the idea that another monkey can see.[132]

Why should it matter so much to understand which sensory modality the other is employing? Because, to repeat, the different sensory systems—vision, hearing, smelling, and so on—have different remits. If you want to know what another creature is likely to be up to, you need to get on their wavelength. Then you'll be able to apply what you know from your own experience on two levels. First, at the level of perception, you'll be able to guess what kind of knowledge the other is acquiring about the external world. With vision, for example, it will be about objects described in terms of colours, shapes, distance, and so on; with touch, objects described in terms of texture, weight, temperature, and so on. Then, separately, at the level of sensation, you'll be able to guess what kind of feelings the other is having about the sensory stimuli impacting its body. With vision, it will be, say, the sensation of phenomenal redness; with touch, the sensation of cold or pain.

Let's note that this will be where your habit of projecting phenomenal properties onto the objects of perception comes into its own. Remember how it emerged in the earlier discussion that when, for example, you project phenomenal redness onto a poppy, you are in effect making a bridge to other sentient beings. You're seeing the poppy as being 'rubropotent'—as having the power to evoke red qualia in another like yourself. If you're a monkey, I guess that when you project phenomenal redness onto the evening sky, you'll be seeing the sky as having the power to make another monkey feel jittery.

*

So, let's return to Bernie. No, I'm not sure he has a theory of mind. Nonetheless, I'm pretty sure he is a natural psychologist who uses himself as a template for predicting the behaviour of others. He knows that I can see, even if not precisely what I can see. He won't steal food from the table when I'm in the room; he won't steal it even when my back is turned; but he will as soon as I leave the room.

He may, possibly, be capable of figuring out how things look from a different perspective than his own. In one experiment, a dog and its human owner were positioned on opposite sides of a barrier and two identical toys were placed on the dog's side of the barrier, only one of which was in view of the human. When the dog was then asked by its owner to 'Fetch!', it almost always fetched the toy they both could see. Other researchers have shown that dogs will signal to a human that they want food that is out of their reach by pointedly looking back and forth from the person to the food.

But here's the surprise. Dogs will still beg help from the human even when they have every reason to believe the person is blind. Anthropologist Florence Gaunet did a study comparing guide dogs, interacting with their blind owners, to pet dogs, interacting with their sighted owners.[133] The dogs were filmed 'asking' their owners for food. In the author's words:

> The guide dogs were just as likely to gaze at the owner, to gaze at the container and to exhibit gaze alternation as were the pet dogs. Guide dogs were not actually sensitive to the fact that their owners were not responding to their gaze signals. The

> results thus suggest that guide dogs did not understand their owner's different attentional state (i.e. that their owners could not recognize the visual signals dogs emit)...Overall. and in other words, the guide dogs did not detect that their owners could not see (them).

Intriguingly, the one thing the guide dogs did do differently was that they smacked their lips loudly, as if to provide a special non-visual cue, but they didn't do this *instead of* giving visual signals.

What are we to make of this? I'd suggest the explanation is that dogs, as natural psychologists, are actually over-reliant on using themselves as a model. They cannot think outside the box of their own experience. The guide dogs have never found themselves with their eyes open in a lighted room and unable to see. So, the state of blindness is just too weird an idea to grasp.

In some circumstances, we humans can be similarly blinkered. We are not at all good at appreciating that other people may have sensory limitations that we ourselves don't have or that other animals don't have. We assume that where sense organs like ours are present, their owner must be using them in the way we are accustomed to. So, for example, we're likely to assume that dolphins, with their large and obvious tongues, must be able to taste. And we'll be surprised to learn that, in fact, dolphins can't taste at all. In this respect, we humans are poor mind-readers because we assume dolphins are more like us than they are. And I expect it goes both ways. Dolphins as natural psychologists probably assume that humans can echolocate like they can. Bernie probably assumes I can smell much better than I can. And I return the favour and assume he has good colour vision—I'm still surprised that he can't find the orange ball in the grass.

*

Does Bernie regard me as sentient? Does he care?

Emmanuel Levinas described a dog who used to wander into the Nazi labour camp where he was imprisoned. 'The dog was always glad to see the prisoners, and thus was the sole creature to treat them as humans. He knew perfectly well that his imprisoned friends were sentient beings and he treated them as such, while their own conspecifics, the Nazi prison guards, did not.'[134]

Everything suggests that Bernie does indeed regard other dogs and humans not just as individual bodies but as individual selves worthy of respect. He greets friends who have been absent as if welcoming back another *mind*. He wants to sit with me while I'm writing as if he's reassured by *my conscious presence*. He's jealous if I show interest in any other dog as if unwilling to share *my affection*. And, at least in some circumstances, he behaves protectively to those, as well as me, whom he counts as family—as if he *feels for us*.

Many species of mammals and birds have been observed to show an empathic concern for the welfare of others in their social group and sometimes for strangers too. In his *Historia Animalium*, Aristotle considered the temper of animals, highlighting the gentleness of the lion, the sensitivity of the elephant, and the kindness of the dolphin who will save his companions from fishermen or support a dead calf to prevent it from being taken by predators. He mentioned in particular dolphins' 'manifestations of passionate attachment to boys' and told stories of dolphins giving young boys rides, pulling them through the water.

Aristotle would not have been surprised to hear of how a female gorilla in the Chicago zoo came to the rescue of a three-year-old boy who had climbed through a barrier and fallen into the gorilla's enclosure: how she gently picked him up and took him to the door where the zookeeper could attend to him.[135] Nor would he have doubted the poet Helen MacDonald's account of how a swan apparently took pity on her when she was sitting mournfully on a riverbank in Cambridge while suffering from a broken love affair. The swan waddled over and sat beside her:

> I watched her, her snaky neck, black eye, her blank hauteur. I expected her to stop, but she did not. She walked right up to where I sat on the step, her head towering over mine. Then she turned around to face the river, shifted left, and plonked herself down, her body parallel with my own, so close her wing-feathers were pressed against my thighs.[136]

When an animal gives comfort to another individual who has suffered a defeat, as in a way MacDonald had, ethologists call it 'consolation behaviour'. The context is usually physical aggression. Individuals fight over dominance, over mates, over food. One wins, one loses. But in many species, from chimpanzees to wolves, to rooks, bystanders who have seen how the fight unfolds, rather than rushing to ingratiate themselves with the winner, typically offer support and reassurance to the loser by bodily touching or grooming.

The urge to come to the aid of another individual in trouble has been studied experimentally in a variety of animals. In an experiment with dogs, the dog's owner sat in a room behind a closed door that the dog in the neighbouring room could open by pressing with its nose. The owner either cried as if in distress or hummed a tune. The dogs opened the door much more quickly when the owner cried.[137]

In a more dramatic study with rats, the rat was able to take action to help another who appeared to be at risk of drowning. There was a box with two compartments divided by a transparent partition. On one side of the box, a rat was forced to swim in a pool of water, which it strongly disliked. On the other side was a rat on a dry floor. The dry rat could rescue the swimming rat by pushing open a small door between the two sides, letting it climb through onto dry land. Within a few days, the dry rats quickly learned to aid their soaking companions by opening the door. They did not open the door when the pool was dry, confirming that they were helping in response to the other's distress rather than because they wanted company. Furthermore, it was found that when the roles were reversed, rats that had themselves had the experience of being soaked learned how to save their cage-mates much more quickly than naive rats, suggesting that as natural psychologists they were moved to action because they knew what it was like.[138]

In these experiments, animals have been given the opportunity to take positive action to protect another in distress. But equally significant are experiments where they could, if they chose, *avoid* taking action that would *cause* distress. Sixty years ago, a cruel experiment was done with rhesus monkeys. The monkeys were taught to pull a chain to dispense a food pellet: pulling one chain in response to a red light and the other in response to a blue light. For three days, the monkeys performed the task happily enough. Then, on the fourth day, one of the chains was programmed to administer a high-frequency electric shock to a monkey in the chamber next door, visible through a one-way mirror. What happened next was remarkable. From then on, the majority of the monkeys refused to pull the chain that gave the shock. They would rather go hungry than hurt the monkey next door. But, again, personal experience was critical. When the roles were reversed, monkeys who had previous experience of being on the receiving end of the shocks were particularly reluctant to do it.[139]

Evidence like this might seem to suggest that mind-reading and compassion inevitably go together. Perhaps any sentient creature who feels any another's pain will be driven by fellow feeling to want to help reduce it. It would be wonderful if this were the general rule. But even the most sentimental of us must know it isn't. It's not always true of other animals we can presume to be sentient, and I have to say in particular it's not always true of Bernie. While Bernie will show tender concern for those in his family group, he can sometimes behave atrociously to those outside it. He terrorizes the cat. And the postman. But worse, he will hunt down and kill other sentient creatures to which he has no family connection. Recently, I heard screams coming from the bottom of the garden. Bernie had trapped a muntjac deer against the fence and was viciously biting it, undeterred by its pitiful cries. A younger deer—its offspring?—was watching transfixed.

I confess I found this incident both worrying and puzzling. Bernie recognizes that when I cry I'm in pain. So, surely he recognizes that when the deer cries it's in pain. He cares for me when I'm in pain. Then why doesn't he care for the deer? The conditionals would seem to be mounting up. 'If you're in pain and you're one of my own, I'll take care of you; but if you're in pain and you're a deer, I'll carry on biting you regardless.'

There was another aspect to this situation that I found puzzling. Why did the deer cry? What was the point? Presumably, animals have evolved to cry when they are under attack either to call for help from others of their kind, to warn others to keep clear, or else to threaten the attacker. But in this case, there was no way that another deer could have helped. The young one with it didn't run away. So, I can only think the purpose of crying was to disorient or disconcert the dog—in which it clearly failed.[140]

What's all too clear is that the fellow feeling Bernie shows for other sentient beings is highly selective. And, of course, it's not just Bernie. No animals are universally compassionate. I've been shocked to see chimpanzees, who can be as tender to each other as any dog, eating a wailing colobus monkey alive. And worse, beating up and killing another chimpanzee who they have known all his life as one of their own.

Then how about humans themselves? Charles Darwin wrote:

> Sympathy beyond the confines of man, that is humanity to the lower animals, seems to be one of the latest moral acquisitions. It is apparently unfelt by savages, except towards their pets. How little the old Romans knew of it is shewn by their abhorrent gladiatorial exhibitions. The very idea of humanity, as far as I could observe, was new to most of the Gauchos of the Pampas.[141]

The fact is, with humans compassion is as selective as it is for dogs and chimpanzees. Humans can care genuinely and deeply for some sentient creatures while they trample on others.

In a letter to *The Spectator* in 1945, the Nobel Laureate A. V. Hill drew attention to the Nazis' unexpected concern for animal welfare:

> One of the first legislative acts of the Third Reich was to issue an Animal Protection Law dated 24 November 1933, and signed by Hitler himself. The following details of it supply an ironic comment on recent revelations of Nazi cruelty. Section I stated:
>
> 1. It shall be prohibited unnecessarily to torture or brutally to ill-treat an animal.
>
> 2. To torture an animal is to cause it prolonged or repeated pain or suffering; the pain inflicted is deemed unnecessary when it serves no reasonably justifiable purpose. To ill-treat an animal means to cause it pain. Ill-treatment is deemed brutal when it is inspired by a lack of feeling.
>
> Among the prohibitions of Section II were the following small-scale models perhaps of the Nazi treatment of Jews, political opponents, foreign workers, and prisoners-of-war:

1. By neglect, to inflict pain or injury in the maintenance, care, housing, or transport of animals.

2. To use an animal wantonly for the performance of work which is obviously beyond its strength, or which is calculated to cause it pain, or for which its condition renders it unfit.

3. To abandon one's own domestic animal with the object of getting rid of it.

4. To sharpen or test the keenness of dogs by using cats, foxcubs, or other animals for the purpose.

In Section III strict regulation was provided of the use of living animals for purposes of research. Göring was a lover of dogs and may have induced his master to lump scientific research and cruelty to animals together.

In Section IV severe penalties of fine and imprisonment were prescribed for torturing or ill-treating an animal, or for performing experiments on living animals for purposes of research without the necessary licence. If, under German law, men may claim the same rights as animals, then tens of thousands of Nazi criminals could be severely punished under Hitler's own Animal Protection Law of 1933.[142]

Savages, except towards their pets!

The passage by Darwin continues: 'This virtue [of kindness to lower animals], one of the noblest with which man is endowed, seems to arise incidentally from our sympathies becoming more tender and more widely diffused, until they are extended to all sentient beings.' But the extension of human sympathies to all sentient beings—when and where it's happened—is, as he recognized, 'one of the latest moral acquisitions'. It's a cultural trait that will not have been driven by natural selection. And it remains disturbingly open to cultural revision.

So, to the questions 'Does Bernie regard me as sentient?' and 'Does he care?', I think I can safely say the answers are 'Yes'. But

I should add that the reason he cares is that he cares about *me* being sentient, not because he cares about sentience in general. If the questions were 'Do I regard Bernie as sentient?' and 'Do I care?', the answers would be 'Yes' again. But, if I'm honest, the same proviso would apply.

Nonetheless, the fact that we both care—even if the caring is conditional—is proof positive that we start from the position of being sentient ourselves.

A final question: what does—or could—Bernie know about death and the extinction of sentience?

Near to where my family had a holiday house in Ireland, there lived a terrier dog, Jack, and his elderly owner, Tom. We would often see them out for walks by the lake, the dog trotting along at the heels of his master. When the old man died of a heart attack, the dog was taken in by neighbours. For the next two years, Jack returned every day, come wind or weather, and sat in the lane outside his former home. He was sitting there when he was hit by a car and killed.

On the night Tom died, Jack had seen him collapse. He had licked his face as he lay on the floor. He had seen the ambulance take the body away. He must have known that something was amiss. As the days passed and Tom didn't show up, it must have become all too clear to the dog that his owner was no longer behaving like his usual self. But the one thing that would never have occurred to Jack was that Tom's self had completely vanished, that in the body he had seen taken away there was now no one at home—no one to put the body on its feet again or to remember Jack's licks.

Jack, I assume, was a natural psychologist like Bernie. And the possibility of permanent oblivion is something for which a natural psychologist is ill prepared. As you look at a lifeless body, you will not find anything in your own experience to match it. You have never had first-hand experience of being dead. The nearest thing you know of is probably sleep. But sleep, although it does involve temporary oblivion, is a misleading model for death: it's a state from which you always awake.

We shouldn't be surprised therefore if non-human animals don't get the meaning of death. They are bound to be puzzled, confused, even angered by the turn of events. They may indeed sometimes confuse it with sleep and expect that they can shake the deceased out of it. An extraordinary video shows a rhesus monkey trying to revive another who had been electrocuted while walking on high-tension wires above a railway in India. The monkey was knocked unconscious and fell onto the tracks. His companion picked him up, slapped his face, and doused him repeatedly in a water butt.[143]

Guess what. In this case, astonishingly, the electrocuted monkey revived. It seems that the adage 'Never say never' does occasionally pay off. Perhaps it pays off just often enough to explain other examples of animals tending to the dead. When the dolphins kept their dead calf afloat, perhaps it was not, as Aristotle claimed, to prevent it from being eaten but rather in the hope it would wake up. When elephants tend to the bones of a long dead family member, perhaps they are still half-expecting the bones to get up and put on their skin again.

In many ways, humans have no better understanding of death than does Bernie. But in one big way, humans are much better informed about the facts. Their ability to retain and pass on knowledge within the human community means they know full well that dead bodies don't come back to life again, that everyone dies eventually, and, most importantly, that it's going to happen to them. They know for a surety that it's not equivalent to sleep and they have every reason to believe that oblivion will be permanent. *Unless*—and here's the wonderful possibility—the self can escape into an afterlife.

The belief, so widely held by humans, that the self can survive bodily death and continue its existence in an alternative world is a bold and brilliant conjecture. It has several things going for it as a successful meme.

It's *common sense*. Everything you know from first-hand experience speaks to the staying-power of the self. If your self ever goes missing, as it does sometimes, it is able to reboot. It's as if self is conserved, almost in the way that matter is conserved. It stands to reason, therefore, that when and if your self is no longer attached to your body here on Earth, it must exist in disembodied form somewhere else—in heaven, elysium, valhalla, or wherever the home of the ancestors is.

It's *comforting*. If something matters, it matters. The self whose existence you value so highly while you live will not cease to matter just because you're dead. Thankfully, in the next world your self can continue to matter to you and others for ever and ever. True, there are cynics, like Professor Broad whom we met earlier, who claim not to want this: 'For my own part I should be more annoyed than surprised if I should find myself in some sense persisting immediately after the death of my present body.' But we needn't take Broad's word for it: I doubt this annoyance, if tested, will have lasted.

It's *irrefutable*. Nothing on Earth can prove the belief in survival to be wrong. Meanwhile, there are enough sporadic examples of the selves of the dead apparently interfering in the affairs of the living to suggest that it has to be right—prayers answered, spirit communications, ghostly visitations. Even if such events are rare and not given to everyone to witness, the evidential base for the belief in survival grows stronger every time.

It is hardly surprising, then, that the belief did in fact become embedded in the minds of humans early on—probably as soon as people could debate it among themselves. We know that humans started burying their dead with grave goods some 50,000 years ago. But we can safely assume that the idea took hold long before that. And, ever since, it has been providing anxious humans with, on the one hand, an antidote to existential despair and, on the other, a powerful incentive to conduct themselves commendably in order to secure the approval of the dead who may be watching. This has undoubtedly had tangible benefits for both the individual and the community. There will therefore have been selective advantage to whatever psychological traits might help to make the belief in an afterlife stick.

Indeed, I see a perfect storm as having arisen in the later stages of human evolution. By 100,000 years ago, humans were already psychologically unique in several ways: they were exceptionally sensual as well as sapient, they had high self-esteem, a sophisticated theory of mind, wide-ranging compassion—and they were on the threshold of developing a linguistic culture that could curate ideas about the soul, death, and survival. But underlying every aspect of this package was the phenomenal self, created by the qualities of bodily sensations. Because the package promoted human fitness, anything that could make the self more remarkable and more convincing was going to be hungrily adopted by natural selection. I suggest that such was the very special context for the elevation of phenomenal consciousness to the extraordinary size and shape that humans know today.[144]

22

TAKING STOCK

I confess the thought has crossed my mind that humans could be the *only* sentient animals on Earth: that Descartes could have been right all along in claiming that non-human animals are unconscious machines, that Dennett could be right that human language could be making such a difference to the quality of sensory experience that no animal experience is comparable. However, when I look at everything we've been discussing, the heretical thought vanishes. While I think we should accept—and even embrace—the idea that humans are both more sentient and more cognisant of it than any other animals, we have already adduced reasons enough to be sure it isn't only humans. Chief among the reasons is the kind of evidence we've examined in the last two chapters.

Admittedly, the positive evidence for sentience in non-humans is patchy. It may not be all we might have hoped for. But it's there. And we mustn't let the bright light of human sentience diminish the strong glow we've detected coming from other species. Some species have stood out in our survey—chimpanzees, dogs, parrots. Some have got an honourable mention—rats, dolphins, monkeys. Most species, it's true, haven't featured at all. Would hedgehogs pass any of the tests? Would ostriches? Would dogfish? Without specific empirical studies, we can't know for sure.

But we're evolutionists. And the fact that we have good evidence for at least some mammals and some birds means we can make an informed guess about others based on taxonomy. Animals that share a recent common ancestor with the species for which we do have convincing evidence are very likely to be sentient.

In Chapter 16, before we came to the evidence, I argued, on theoretical grounds, that phenomenal consciousness is restricted to warm-blooded animals. It wasn't a physiological possibility until the brain warmed up, nor was it ecologically relevant until animals became relatively free of environmental constraints.

If that's right, the simplest evolutionary scenario would be that sentience arose early on in the stems of both mammals and birds and thereafter has remained a universal feature across their descendants. This would imply that, if we can be sure that *any* mammal or *any* bird is sentient, we can be pretty sure *all* are: that's to say all 6,000 mammal species and all 10,000 bird species. But there's a plausible alternative scenario. This would be that sentience has actually arisen several times among mammals and birds, not in the stems of the evolutionary tree but later on, in the branches to which they gave rise.

As we discussed earlier, the upgrading of sensations would have been a relatively easy step for evolution once conditions were ripe: that's to say, once the brain was prepared for it and there were benefits to be had from a phenomenal self. But the second condition may not have been met immediately, the earliest mammals and birds may not have been up for living in a society of selves. In that case, some branches going forward might have missed out on sentience altogether and have remained insentient to this day.

For what it's worth, I'm inclined to believe in the second scenario. I wouldn't be surprised if it were to turn out that there are still odd-ball species of mammals and birds that pass none of our tests. It won't be *all* mammals and birds that are sentient. Just most of them.[145]

*

Whenever it occurred, sentience had to *start* somewhere. The advent of warm-bloodedness neatly provides for there being a defining moment. As Dennett has summarized my view of this:

> There was a great bifurcation in evolution: the warm-blooded mammals and birds literally had the time and energy to escape into the rarefied design space of phenomenal consciousness, while the rest of the living world had to settle for varieties of zombie cleverness. This is certainly a startling idea: if Humphrey's right, there is nothing *it is like* to be an octopus (in spite of their beguiling behavior) but there is something it is like to be a chicken.[146]

For many people, it's too startling. The objection I've heard over and again is precisely that *it leaves out octopuses.*

The question of octopus sentience, hardly an issue for most people in the past, has been brought centre stage by new scientific findings and by popular accounts of human encounters with these alien creatures.[147]

The thinker who has done most to draw attention to the possibility that octopuses are conscious, if not in the same way we are, at least in a way we should recognize and respect, is Peter Godfrey-Smith, a philosopher and a naturalist who has made extensive observations of octopuses in the seas around Australia.

Godfrey-Smith is an enthusiast. He sees octopuses as marvels of biological design. But he's a scientific sceptic too. He approaches creatures so different in brain and bodies from human, knowing that human intuitions may well be a poor guide. In fact, he counsels against reading too much into examples of their apparent 'braininess':

> People often now talk about octopuses as 'smart', and in some ways they are. But that is not the term that comes readily to my mind...Octopuses are exploratory animals who direct the complexity of their bodies on whatever confronts them. They fiddle about and try things and turn the problem over and over—physically, not mentally...They are not, for the most part, ruminative and 'clever' sorts of animals.[148]

Nonetheless, while playing down exaggerated claims about octopuses' general intelligence, he picks out several areas of behaviour that chime with our tests of sentience. In particular, he suggests that octopuses are playful, social, and psychologically astute.

But are they? If you're hoping he means that octopuses are really quite like dogs in these respects, you'll be disappointed.

As evidence of play, he cites their interest in exploring novel objects. Yet, there's nothing to suggest that they seek out sensory experience for the sake of enlarging what they know about their own capacity for feeling rather than their knowledge of the world out there. As evidence for sociality, he cites a female octopus throwing debris at a male, with the apparent intention of getting him to back off. Yet, there's no evidence that octopuses ever collaborate or form intimate relationships with others. As evidence of mind-reading, he cites anecdotal evidence that an octopus may take account of whether a human can see it when it goes into hiding. But of octopuses understanding what it's like to be another octopus there's no evidence at all.

On the face of it, then, octopuses are not qualiaphiliacs, they are not natural psychologists, they don't regard each other as selves, nor do they care. I'm bound to say therefore that the likelihood of their being sentient and having a phenomenal self is negligible.[149]

Godfrey-Smith is actually quite guarded about this. 'Ten years of following octopuses around and watching them...have left me with no real doubt that octopuses experience their lives, that they are conscious in a broad sense of that term.'[150] But this is philosophically tendentious. Does he mean they are cognitively conscious (which I'd say is quite plausible) or phenomenally conscious (which, on the evidence, is quite implausible)?

Disappointingly, in his book *Metazoa,* from which these quotes are taken, he fails to make this distinction and lines up with those theorists who regard phenomenal consciousness as something that simply pops into existence in a complex brain—an intrinsic property *of* brain activity rather than a property of sensations as represented *by* the brain. Indeed, he derides the very idea of qualia as representations. 'Qualia are not extra things that need an explanation, somehow produced by the workings of the physical system. Instead they are part of what it is to be the system.'[151] To my mind, this explains nothing at all.

23

MACHINA EX DEO

A self-driving car, navigating by GPS, can be said to have a point of view. It knows where it is and where it's headed. The car's 'mind' diagnoses potential threats and gives the alarm: low on petrol, overheating, underinflated tyres. It may flag a more serious problem: 'Engine failure. Stop and seek assistance.' And if it detects that there's likely to be a crash, it may slam on the brakes and deploy air-bags.

None of this requires phenomenal consciousness. The car is not sentient. But *could* a machine be sentient?

I've been concerned, in this book, with the possibilities of sentience in living animals that have evolved by natural selection. The theory I've put forward is a theory of how phenomenal consciousness could be generated in a warm brain made of nerve cells. In devising this theory, I've been concerned with how to get a brain to manifest the behaviours and attitudes that go with sentient experience.

Of course, the material of the brain has had to be up to the job (conduction speed had to be fast enough, for instance). But the brain hasn't had to be made of nerve cells. If the theory is right, then in principle a robot with a brain made of silicon or any other suitable material could be designed by human engineers to have equivalent experiences and act in equivalent ways as a result of them.

But the operative word here is 'designed'. In the course of the discussion, we've considered and rejected the idea that phenomenal consciousness could possibly establish itself *unbidden*, simply as a corollary of burgeoning intelligence or ever more complex information processing. Where it has evolved in animals, it has been a relatively late addition, involving dedicated circuitry, selected because of the effects on the animal's psychology. Presumably, if a robot's brain were to contain a module that duplicates the functions of this special circuitry, there's reason to think that the robot would be phenomenally conscious like the animal.

The engineer's task would be to create a *functional* duplicate: a brain that, when the robot's sense organs are stimulated, *does* all the things that a sentient animal's brain does: represents what's happening to it as a sensation with phenomenal qualities, incorporates this into its idea of its self, and has the attitudes and behaviours that follow from this. In short, the end goal would be to design an artificial brain for which the question 'And then what happens?' gets the same answer as in the sentient animal.

It hardly needs saying that this is a tall order and the design of such a brain is not going to happen any time soon. First, the circuits that we've postulated to underwrite phenomenal consciousness in animals will have to be studied in detail, through the combined efforts of neuroscience and theoretical modelling. Then, when we know precisely what these circuits are doing, engineers will have to find a way of programming an artificial brain to do the same thing.

However, let's imagine that fifty years from now engineers, working in a secret research laboratory, announce that they have done it: built a sentient robot. How could we establish, without knowing anything about the robot's internal architecture, whether they have in fact succeeded?

The problem would be very similar to that we've been grappling with in assessing claims about sentience in non-human animals. However, there's one thing might make it considerably more tractable. It will almost certainly be easier to design a robot to understand human language than to design it to be sentient. We can fairly assume therefore that the sentient robot will already have language built in.

With language as a given, the philosophers Susan Schneider and Edwin Turner have proposed a series of conversational tests.[152] They write:

> Each of us can grasp something essential about consciousness, just by introspecting; we can all experience what it feels like, from the inside, to exist.
>
> Based on this essential characteristic of consciousness, we propose a test for machine consciousness, the AI Consciousness Test (ACT), which looks at whether the synthetic minds we create have an experience-based understanding of the way it feels, from the inside, to be conscious.
>
> One of the most compelling indications that normally functioning humans experience consciousness, although this is not often noted, is that nearly every adult can quickly and readily grasp concepts based on this quality of felt consciousness. Such ideas include scenarios like minds switching bodies (as in the film Freaky Friday); life after death (including reincarnation); and minds leaving their bodies (for example, astral projection or ghosts). Whether or not such scenarios have any reality, they would be exceedingly difficult to comprehend for an entity that had no conscious experience whatsoever. It would be like expecting someone who is completely deaf from birth to appreciate a Bach concerto.

> Thus, the ACT would challenge an AI with a series of increasingly demanding natural language interactions to see how quickly and readily it can grasp and use concepts and scenarios based on the internal experiences we associate with consciousness. At the most elementary level we might simply ask the machine if it conceives of itself as anything other than its physical self. At a more advanced level, we might see how it deals with ideas and scenarios such as those mentioned in the previous paragraph. At an advanced level, its ability to reason about and discuss philosophical questions such as 'the hard problem of consciousness' would be evaluated. At the most demanding level, we might see if the machine invents and uses such a consciousness-based concept on its own, without relying on human ideas and inputs.

You'll see why I like these suggestions! But I'd want to add to them. Schneider and Turner, like most other philosophers who speculate about the possibility of conscious robots, take no account of the robot's origins. They don't ask why anyone would have *wanted* to install phenomenal consciousness in a machine. So they don't ask for evidence that it's working in the way that it's designed to. Thus, they fail to emphasize the sensory dimension of phenomenal consciousness and don't think to inquire whether the robot would show evidence of qualiaphilia: whether it would *like* being conscious and would go out of its way to listen to music, for example. But more seriously, nor do they consider how the robot would be able to take advantage of having a phenomenal self in its dealings with others. So, they don't pick up on practical issues such as empathy and mind-reading.

But this risks missing the point. As I see it, the robot's phenomenal self won't just be a secondary feature that happens to accompany sentience as an unexpected bonus. In fact I imagine this will have been the very reason for wanting to build a sentient machine to start with. Why *would* the engineers have wanted to do it? Perhaps it could simply be vanity—to build a machine that resembles themselves. But that's unlikely to have got the project funded! The best reason I can think of is that they will have researched the evolution of sentience in animals, including humans, and have been impressed as we have been by the role the phenomenal self plays in fostering self-esteem and deepening social relationships.

It's the possibility of robots that we humans can relate to—phenomenal self to self—that will have brought the funds in. In the coming decades, robots are likely to become increasingly integrated into the lives of human beings. They are also likely to find themselves interacting with other robots in robot communities. The 'fittest' robots will be those that have a sense of their individuality coupled with the basic skills of a natural psychologist. They will need to be able to go some way to reading the minds of humans and other robots and to be readable by others in their turn.

Robot-to-robot intersubjectivity will presumably become especially important once robots are living in autonomous robot colonies. We humans are already sending robots into space to carry out tasks on our behalf under conditions where humans could not survive. The time is going to come when we'll want to establish self-perpetuating colonies of robots on distant planets, with a mission to build a new life for themselves and only occasional contact with humans back at home.

If they are to make their own way and overcome major intellectual challenges as well as material ones, these robots will have to have inquiring minds. They'll need to be scientifically imaginative and philosophically reflective. But this could prove dangerous. For the more they resemble humans in these respects, the more likely it will be that these robot missionaries could succumb, in their own way, to existential despair—to dark thoughts about whether their lives have any meaning. If that's so, belief in an afterlife for the robot's self—after its body has become defunct—could prove as adaptive for them as it has for human beings.

*

Will this be the whole story of the engineers' ambitions? I can see a different kind of reason—an ethical, not a practical one—why humans may one day be motivated to build sentient robots.

I wrote in the Prologue:

> Suppose conscious beings like us have not evolved anywhere else. Suppose consciousness as it exists on Earth is a one-off accident of evolution... Frank Borman, looking from the window of Apollo 8, remarked 'the Earth is the only thing in the universe that has any colour'. That can't be strictly true. But it could be true that the Earth is the only place where sensations of colour exist. Or sensations of anything: sweetness, warmth, bitterness, pain.

In the pages that followed, we've found several arguments to support this supposition. While we needn't doubt that there are many other life forms out there in the universe, we've come to see that the evolution of life, even intelligent life, will not necessarily have entailed the evolution of phenomenal consciousness. On Earth, it has so happened that a sequence of 'lucky' breaks paved the way for it to evolve as it has done in mammals and birds. On Earth, if the same local conditions were to hold, it's quite possible that the sequence could be repeated. But outside the Earthly environment all bets are off. The chances of phenomenal consciousness having evolved somewhere else in the universe could be vanishingly small.

One day, long in the future as a result of the aging of the sun (or not so long as a result of natural catastrophe or human mismanagement), it's inevitable that life on Earth will become extinct. Looking ahead to that day, it may be some comfort to know that life will carry on elsewhere. It will be rather less comforting, however, if we have reason to suspect that extraterrestrial life will be entirely insentient.

I like to think, therefore, that human beings, in an act of cosmic generosity, will try to forestall the extinction of phenomenal consciousness by seeding the universe with sentient robots.

Thomas Mann, in an essay on 'What I believe', wrote:

> In my deepest soul I hug the supposition that with God's 'Let there be', which summoned the cosmos out of nothing, and with the generation of life from the inorganic, it was man who was ultimately intended, and that with him a great experiment was initiated, the failure of which … would be the failure of creation itself, amounting to its refutation. Whether that be so or not, it would be as well for man to behave as if it were so.[153]

Few of us would want to follow Mann in putting the human species on a pedestal like this. But suppose he had written 'it was phenomenal consciousness that was ultimately intended, and with this a great experiment was initiated, the failure of which would be the failure of creation itself'. I could go with that. Even if the idea of a naturally evolved feature being 'intended' must be wrong, I imagine that Darwin himself could have seen phenomenal consciousness as an 'ultimate' achievement—the crowning glory of the evolutionary process that began with the Big Bang.

It's an invention so sublime that, if it were to cease to exist, it would indeed diminish the whole of creation.

24

ETHICAL IMPERATIVES

Mary Oliver wrote that it would be 'terrible to be wrong' about stones.[154] I agree it would be terrible to be wrong about octopuses, lobsters, locusts, or any other of the world's creatures that our analysis has excluded from the club. I too worry that I may be missing something. I confess I open each new issue of the journal *Animal Sentience* with a frisson of excitement, wondering if I'll find something in it that will throw me.

However, I have to say that at this point I don't expect it. I asked a set of leading questions at the beginning of the book, and I'm ready to take ownership of the answers we've arrived at.

I suggested at the start that it may not be 'self-evident' that humans have a duty of care towards all sentient creatures. But things look different in the light of our discoveries about how phenomenal consciousness has evolved and what it's *for*.

We've seen that it's the job of the phenomenal self to *matter*: to matter in the first instance to its owner but also to other related individuals who have kindred selves. As an evolutionary adaptation, this is how phenomenal consciousness contributes to biological survival. The phenomenal self *exists to be minded about*.

It doesn't follow, of course, that either humans or other sentient animals will automatically mind about the selves of everyone else. In fact, as we've seen, they generally don't;

compassion is selective. Even for humans, sentience doesn't *force* our hands or hearts.

However, human ethics come from somewhere else. We act ethically—if we do—not from *instinct* (which would mean we had no choice) but from *understanding*. We choose to care about the feelings of some but not all non-human animals because—rightly or wrongly—we *recognize them to have selves like ours*. Then we are guided by the foundational principle that has been called the Golden Rule. We consider ourselves obliged to treat these other sentient animals in the way *we would wish to be treated if we were in their place*.

*

In this book, we've explored how far our human beliefs about the extent of sentience are justified. We've undercut some of them, while putting others on a firmer scientific footing. If I said at the start that 'We lack direct evidence and even agreed arguments as to how far consciousness extends', I'm now pleased to withdraw this. The evolutionary theory we've outlined means we are no longer working in the dark.

Many people (probably most) already believe they know pretty much *what it's like* to be a dog—to feel the pain of a thorn, the coldness of a river, the sound of a whistle. Are they right? I'd say we have scientific grounds for saying they *are*. In crucial respects, humans really can enter the dog's consciousness. We have therefore found, in science, compelling reasons why we should extend our ethical concern to dogs. By the same token, we've found reasons why we need not extend the same level of concern to octopuses.

Of course, sentience isn't everything. There are plenty of reasons why humans should care for the welfare of the world's animals irrespective of whether or not they are phenomenally conscious. We should care *for* them even if their not being sentient means there is less reason to care *about* them. We should care for them as part of the web on which life on Earth depends, as miracles of biological design in their own right, as things of beauty, as co-voyagers with humans into the future.

Yet, sentience still counts above all else. Animals that are phenomenally conscious have an absolute claim on us that insentient animals do not. So the welfare of dogs should matter to us more than that of octopuses because dogs *matter to themselves* in a way that octopuses do not. If you were in the place of a sentient dog, you would mind about being treated well by humans; if you were in the place of an insentient octopus, you wouldn't.

This puts us humans under notice to get it right. We must be able to trust that our beliefs about who or what in the world is sentient are based on good intelligence. I'll say it again: terrible to be wrong, but *irresponsible not to be right*. Unjustified beliefs about the extent of sentience can only distort our proper relationship with the natural world and, future manmade worlds as well.[155]

I'm not going to sermonize. When it comes to ethics, science can only propose, while each of us as a thoughtful individual disposes.

So, over to you. No one knows better. I'm not going to end this book on a down note.

*

I once wrote a review of anthropologist Gregory Bateson's book, *Mind and Nature*.[156] Bateson was maintaining that the natural world is one vast mind and that we, as human beings, should treat nature with the respect due to a conscious being. I remarked that 'There are, I know, good reasons for not cutting down the Amazonian forests: but the idea that such destruction is equivalent to psycho-surgery is not one of them.'

Bateson replied in the letter column of the paper, accusing me of putting logic ahead of poetry. The following argument, he said, is bad logic, but it's good poetry: *Men die. Grass dies. Men are grass.* I replied, in turn, that the following argument is bad logic and bad poetry. *Men die. Turnips die. Men are turnips.*

Bateson sent me a nice postcard: 'Dear Nick, *Touché.* I hope we are still friends. As one turnip to another. Yours, G.B.'

Philosophers—and especially students of consciousness—need to have a sense of humour.

ACKNOWLEDGEMENTS

This book has been some time in the making, and more people than I can say have supported me in getting to this point. Dan Dennett and I have been in regular—often daily—contact over the years. Others have read and advised me on parts of the text: notably Paul Broks, Tom Clark, Keith Frankish, Sam Humphrey, Geoffrey Lloyd, Chris McManus, Michael Proulx, Nick Romeo, and Chris Sykes. My agent, Toby Mundy, encouraged me to write the book and has stood by me throughout. Latha Menon at Oxford University Press and Philip Laughlin at MIT Press have lovingly shepherded it into print.

REFERENCES AND NOTES

1. William Youatt (1839). *The Obligation and Extent of Humanity to Brutes, Principally Considered with Reference to the Domesticated Animals*, repr. 2003, intro. R. Preece (Lewiston, New York: Edwin Mellen Press).
2. David Chalmers (2018). 'How Can We Solve the Meta-Problem of Consciousness?', *Journal of Consciousness Studies*, 25, 6–61, 6.
3. Ned Block (1995). ''On a Confusion about a Function of Consciousness', *Behavioral and Brain Sciences*, 18, 227–247.
4. Sleep-texting too. In a survey of American college students, it was found that 25 per cent had sent semi-coherent texts on their mobile phones while asleep. Elizabeth B. Dowdell and Brianne Q. Clayton (2019). 'Interrupted Sleep: College Students Sleeping with Technology', *Journal of American College Health*, 67,7, 640–646.
5. Jeremy Bentham (1789). *Introduction to the Principles of Morals and Legislation*, Chapter 17. The complete passage reads:

   > The day may come, when the rest of the animal creation may acquire those rights which never could have been withholden from them but by the hand of tyranny. The French have already discovered that the blackness of skin is no reason why a human being should be abandoned without redress to the caprice of a tormentor. It may come one day to be recognized, that the number of legs, the villosity of the skin, or the termination of the os sacrum, are reasons equally insufficient for abandoning a sensitive being to the same fate. What else is it that should trace the insuperable line? Is it the faculty of reason, or perhaps, the faculty for discourse? ... the question is not, Can they reason? nor, Can they talk? but, Can they suffer? Why should the law refuse its protection to any sensitive being? ... The time will come when humanity will extend its mantle over everything which breathes.

6. Daniel Dennett (2007). 'A Daring Reconnaissance of Red Territory', *Brain*, 130, 593–595, 594.
7. Laurence Klotz (2005). 'How (Not) to Communicate New Scientific Information: A Memoir of the Famous Brindley Lecture', *BJU International*, 96, 956–957.
8. Isaac Newton (1665). *Of Colours*, Laboratory Notebook, 1665, Cambridge University Library, MS Add. 3975, pp. 1–22 (published online 2003).

9. Thomas Reid (1785/1969). *Essays on the Intellectual Powers of Man*, Part II, Ch. 17 (Cambridge MA: MIT Press), p. 265.
10. C. D. Broad (1962). *Lectures on Psychical Research* (London: Routledge and Kegan Paul), p. 430.
11. Hugh Whitaker (1959). *The Eternal Resurrection: The Spiritual Teachings of Agresara*, 3 vols (London: Sidgwick and Jackson).
12. C. D. Broad (1925). *The Mind and Its Place in Nature* (London: Kegan Paul).
13. Ibid. pp. viii, 227.
14. Nicholas Humphrey (1986). 'Is There Anybody There?', Channel 4 TV, 1986, https://www.youtube.com/watch?v=qdOWChIXgd8 (accessed 10 May 2022).
15. Lawrence Weiskrantz (1963). 'Contour Discrimination in a Young Monkey with Striate Cortex Ablation', *Neuropsychologia*, 1, 145–164, 159.
16. Nicholas Humphrey (1968). 'Responses to Visual Stimuli of Single Units in the Superior Colliculus of Rats and Monkeys', *Experimental Neurology*, 20, 312–340.
17. J. Y. Lettvin , H. R. Maturana , W. S. McCulloch, and W. H. Pitts (1959). 'What the Frog's Eye Tells the Frog's Brain', *Proceedings of the IRE*, 47, 1940–1951.
18. N. K. Humphrey and L. Weiskrantz (1967). 'Vision in Monkeys after Removal of the Striate Cortex', *Nature*, 215, 595–597.
19. Nicholas Humphrey, (1974). "Vision in a monkey without striate cortex: a case study," *Perception*, 3, 241–255. There's a YouTube film of Helen: https://www.youtube.com/watch?v=6ek2LBqM7dk. On one occasion Dan Dennett and I showed this film, without a preamble, at a philosophy seminar in Columbia University. We then asked the audience to suggest what if anything might be different about the monkey. No one had a clue.
20. Nicholas Humphrey (1972). 'Seeing and Nothingness', *New Scientist*, 53, 682–684.
21. Lawrence Weiskrantz (1986). *Blindsight: A Case Study and Implications* (Oxford: Clarendon).
22. Beatrice de Gelder, Marco Tamietto, Geert van Boxtel, et al. (2008). 'Intact Navigation Skills after Bilateral Loss of Striate Cortex', *Current Biology*, 18, R1128–R1129.

23. YouTube film of TN: https://www.youtube.com/watch?v=ACkxe_5Ubq8 (accessed 10 May 2022).
24. But if blindsight is a case of pure perception in the absence of sensation, why isn't it *cognitively conscious*—available to introspection—in the way perception usually is? It's an important question. I'm committed to the view that blindsight should indeed be consciously accessible (although the subject may be confused about this because the phenomenal dimension is missing.) As I mentioned above, I had evidence that Helen did indeed know what she was seeing. When she was sitting in the tree, she would reach for a titbit when she could perceive it was within arm's length but ignore it if it was too far away. (You can see an example of this at the very end of the video that is available online (n 19). Helen can be seen to look attentively at a peanut that is almost out of reach: she hesitates, decides not to go for it, but then changes her mind.) With the first human cases of blindsight, Weiskrantz didn't find evidence of any such introspective awareness. But as research has continued and researchers have learned better what to look out for, it has been found with several patients that, although the patient will indeed maintain that he doesn't see the stimulus, he 'is aware' in a vague way of what's out there. This has been called type-2 blindsight. Fiona MacPherson (2015) provides a thorough discussion of the evidence, 'The Structure of Experience, the Nature of the Visual, and Type 2 Blindsight', *Consciousness and Cognition*, 32, 104–128.
25. Thomas Reid (1764). *An Inquiry into the Human Mind*, Ch. 6, 'Of Seeing', section 21, quoted in Ryan Nichols (2007). *Thomas Reid's Theory of Perception* (Oxford: Oxford University Press), p. 155.
26. Thomas Reid, Letter to Lord Kames, quoted in Ryan Nichols, *Thomas Reid's Theory of Perception* (Oxford: Oxford University Press), p. 152.
27. Thomas Reid (1785/1969). *Essays on the Intellectual Powers of Man*, Part II, Ch. 17 (Cambridge MA: MIT Press), p. 265.
28. J. Y. Lettvin, H. R. Maturana, W. S. McCulloch, and W. H. Pitts (1959). 'What the Frog's Eye Tells the Frog's Brain', *Proceedings of the IRE*, 47, 1940–1951, 1951.
29. Carol Ackroyd, Nicholas Humphrey, and Elizabeth Warrington (1974). 'Lasting Effects of Early Blindness: A Case Study', *Quarterly*

Journal Experimental Psychology, 26, 114–124. I have added details here that we withheld in the published paper.

30. Neuropsychologist Paul Broks (personal communication) has suggested a parallel between the partial loss of self that accompanies blindsight and the much more drastic loss that occurs in patients with Cotard's syndrome, who will flatly declare that they are *dead*. In his 2018 book, *The Darker the Night the Brighter the Stars* (London: Allen Lane), Broks describes a case he examined:

    > 'What makes you think you're dead?' 'It's because I'm nothing now. I don't exist anymore.'... Cotard's syndrome is a disturbance of self-awareness whereby the normal intuitions of embodiment and in-the-moment consciousness seem to be severely undermined... In Cotard's, there is a dissolution of the self of the present moment, dissolution to the point of experienced non-existence. pp. 140–144

    Do these patients have disturbances in phenomenal awareness? It's hard to know. But it could be significant that, in the one patient who has undergone a brain scan, there was a marked reduction in the activity of the cerebral cortex, while subcortical areas remained normal.
31. Nicholas Humphrey (1971). 'Colour and Brightness Preferences in Monkeys', *Nature*, 229, 615–617.
32. Richard Passingham (2018). Speech at Memorial for Larry Weiskrantz in Oxford, 8 June 2018.
33. Nicholas Humphrey and Graham Keeble (1978). 'Effects of Red Light and Loud Noise on the Rate at Which Monkeys Sample Their Sensory Environment', *Perception*, 7, 343–348.
34. Colin Groves and Nicholas Humphrey (1973). 'Asymmetry in Gorilla Skulls: Evidence of Lateralised Brain Function?', *Nature*, 244, 53–54.
35. Dian Fossey's darker side is documented in Harold Hayes' biography (1991), *The Dark Romance of Dian Fossey* (London: Chatto & Windus). Hayes describes multiple examples of her deranged behaviour and acts of cruelty, obtained from witnesses.
36. Nicholas Humphrey (1976). 'The Social Function of Intellect', in *Growing Points in Ethology*, ed. P. P. G. Bateson and R. A. Hinde (Cambridge: Cambridge University Press), pp. 209, 303–317.
37. Although 'Dunbar's number' has been enthusiastically endorsed by journalists and politicians, evolutionary psychologists generally

take it with a pinch of salt. Several critics have accused Dunbar of massaging the data to get the result he wants. See, e.g. P. Lindenfors, A. Wartel, and J. Lind (2021). 'Dunbar's Number Deconstructed', *Biology Letters*, 17, 20210158, 2021.

38. Nicholas Humphrey (1980). 'Nature's Psychologists', in *Consciousness and the Physical World*, ed. B. Josephson and V. Ramachandran, (Oxford: Pergamon), pp. 57–75, 73.
39. Keith Frankish (2016). 'Illusionism as a Theory of Consciousness', *Journal of Consciousness Studies*, 23, 11–39.
40. I first used this analogy in Nicholas Humphrey (2008). 'Getting the Measure of Consciousness', in *What is Life? The Next 100 Years of Yukawa's Dream*, ed. M. Murase and I. Tsuda, *Progress of Theoretical Physics Supplement*, 173, 264–269.
41. I have suggested that we might do better to call phenomenal properties 'surreal':

> Neither Illusionism nor Realism addresses what should be the central question for a theory of consciousness: namely, how we represent the meaningful relationship we to have to sensory stimulation…I have a suggestion: Phenomenal Surrealism—where 'surreal' has the meaning that Picasso originally gave it. 'What I intended when I invented this word, [was] something more real than reality'…'Resemblance is what I am after—a resemblance deeper and more real than the real, that is what constitutes the sur-real.' It was in this spirit that Picasso could say of his great sculpture of a goat, 'She's more like a goat than a real goat, don't you think.' My thought, then, is this: just as Picasso's goat was goatier than a real goat, so phenomenal redness is redder than real red, phenomenal pain painier than real pain. In general phenomenal properties are represented in sensation as 'more real' than the actual physiological events that give rise to them. By adding in the relational dimension of how we feel about it, sensation has, as it were, put one over on the physical reality of stimulation.

Nicholas Humphrey (2016). 'Redder than Red: Illusionism or Phenomenal Surrealism', *Journal of Consciousness Studies*, 23, 116–123.

42. Oscar Wilde (1905/1950). *De Profundis: The Complete Text*, ed. Vyvyan Holland (New York: Philosophical Library), p. 104.
43. David Hume (1739). *A Treatise of Human Nature*, Book I, Part III, section XIV.
44. Thomas Reid (1785/1969). *Essays on the Intellectual Powers of Man*, Part II, Ch. 17 (Cambridge MA: MIT Press), p. 265.

45. Sargy Mann, quoted in Peter Mann and Sargy Mann (2008), *Sargy Mann: Probably the Best Blind Painter in Peckham* (London: SP Books), p. 203.
46. Robert Browning (1885). 'Fra Lippo Lippi', lines 300–306.
47. Dan Lloyd (1990). *Radiant Cool* (Boston, MA: Bradford Books), p. 16.
48. René Descartes (1641/1986). *Meditations on First Philosophy*, trans. John Cottingham (Cambridge: Cambridge University Press), p. 35.
49. Colin McGinn (1993). 'Consciousness and Cosmology: Hyperdualism Ventilated', in *Consciousness*, ed. M. Davies and G. W. Humphrey (Oxford: Blackwell), p. 155.
50. Alfred Russel Wallace (1869/2009). 'The Limits of Natural Selection as Applied to Man', in *Contributions to the Theory of Natural Selection* (Cambridge: Cambridge University Press), p. 361.
51. Philip Goff (2019). *Galileo's Error* (London: Rider), p. 21.
52. Bertrand Russell (1919). *Introduction to Mathematical Philosophy* (London: Allen and Unwin), p. 71.
53. Tom Clark (2022, in press). 'Content: A Possible Key to Consciousness' provides detailed arguments to support the case that mental representations can be about the real properties of things that exist *only* at the level of representational content:

    Content items such as concepts, propositions, beliefs, numbers, and, I would suggest, experiential qualities, are terms of representation that get activated in, and deployed by, mind systems like ourselves, but these terms aren't locatable in the spatio-temporal manifold they participate in representing, so aren't perspicuously characterized as physically objective. We don't find concepts, numbers, beliefs, or propositions looking in the head, nor out in the world as delimitable objects, yet they are indispensably real as representational elements used in conceptually modelling reality.

    See also Tom W. Clark (2019). 'Locating Consciousness: Why Experience Can't Be Objectified', *Journal of Consciousness Studies*, 26, 60–85. Clark's treatment is reassuringly close to mine.
54. Daniel Dennett (1991). *Consciousness Explained* (Boston: Little Brown), p. 255.
55. Knock apparition, witnesses statements 1879, https://www.knock-shrine.ie/history/ (accessed 20 May 2022).
56. T. H. Huxley (1870). 'Has a Frog a Soul, and of What Nature Is That Soul, Supposing It to Exist?', Metaphysical Society (8 November).

57. Bjorn Merker (2007). 'Consciousness without a Cerebral Cortex: A Challenge for Neuroscience and Medicine', *Behavioral and Brain Sciences*, 30, 63–134.
58. Mark Solms (2019). 'The Hard Problem of Consciousness and the Free Energy Principle', *Frontiers in Psychology*, 30 January, 9, 2714. doi: 10.3389/fpsyg.2018.02714, 5.
59. Joe Simpson (1988). *Touching the Void* (London: Jonathan Cape), p. 109.
60. Frank Jackson (1986). 'What Mary Didn't Know', *Journal of Philosophy*, 83, 291–295.
61. Robert Burns, (1791). 'Tam o'Shanter', line 231.
62. Samuel Taylor Coleridge (1817/2002). *Biographia Literaria*, in *Opus Maximum: Collected Works*, Vol. 15, ed. Thomas McFarland with Nicholas Halmi (Princeton, NJ: Princeton University Press), p. 132.
63. Barbara Montero, in a ground-breaking essay about 'qualitative memory', discusses the particular case of pain. She demonstrates that, contrary to what is generally believed, we do not retain a rich representation of what it was like to be in pain. However, she explicitly contrasts this with our ability to remember what it was like to see red. I think the two cases are actually no different. See Barbara Montero (2020). 'What Experience Doesn't Teach: Pain Amnesia and a New Paradigm for Memory Research', *Journal of Consciousness Studies*, 27, 102–125.
64. John Locke (1690/1975). *An Essay Concerning Human Understanding*, ed. P. Nidditch, Book II, Ch. XXXII, section 15 (Oxford: Clarendon Press).
65. Could it show up in colour preferences? Most humans have similar preferences if tested under standard conditions. Yet, not all do. Chris McManus, in a study of people's preferences for coloured cards, involving fifty-four people, each making 256 paired comparisons, found that while 70 per cent of his subjects consistently preferred blue/green hues to yellow/red hues, a significant subgroup of 20 per cent showed a distinctly different pattern, consistently preferring yellow/red to blue/green. I. C. McManus, Amanda L. Jones, and Jill Cottrell (1981). 'The Aesthetics of Colour', *Perception*, 10, 651–666.
66. I have expanded on the possibilities of individual differences in phenomenal consciousness in Nicholas Humphrey (2020).

'Consciousness: Knowing the Unknowable', *Social Research*, 87, 157–170.

67. Charles Darwin (1859). 'Difficulties of the Theory', in *On the Origin of Species*, Ch 6, p. 143 (London: John Murray).
68. Giorgio Vallortigara has taken up—and run with—the idea of an efference copy of sentition as the basis for grounding phenomenal consciousness and the sense of self. He has added significantly to my own treatment. Giorgio Vallortigara (2021). 'The Rose and the Fly. A Conjecture on the Origin of Consciousness', *Biochemical and Biophysical Research Communications*, 564, 170–174.
69. Suppose that each time the activity cycles around the feedback loop, the transmission characteristics are altered by this very activity. The growth of the activity in such a circuit will be governed by what is called 'a delay differential equation'. This is an equation where the evolution of the system at a certain time, **t**, say, depends on the state of the system at an earlier time, **t-T**, say. What happens then is that the activity, once started, if it does not quickly die away, will either develop chaotically or soon settle into a 'basin of attraction' in which the same pattern repeats itself indefinitely and to which it returns even if disturbed.
70. It may be that evolution actually took the same path several times in the history of a single species. How else to explain that, for humans, and presumably for other sentient species also, each of the sensory modalities gives rise to sensations with phenomenal properties? Why do we have visual qualia, auditory qualia, smell qualia, and so on? Given that the different sense organs and their brain pathways have long been anatomically separate, this might suggest that each modality had to acquired phenomenal properties independently of the others—but under the influence of the same evolutionary dynamic we've described. However, there's another possibility. Perhaps the genes responsible could be operating at an early stage of embryological development that is common to all the sensory modalities. In that case, the phenomenalization of one modality, whichever was first to be selected, might have brought the other modalities in its train.
71. David Hume (1739/1978). 'Of Personal Identity', in *A Treatise of Human Nature*, section 6, ed. A. Selby-Bigge (Oxford: Oxford University Press).

72. Paul Klee (1920). *Creative Credo*, https://arthistoryproject.com/artists/paul-klee/creative-credo/ (accessed 10 May 2022).
73. Pablo Picasso (1923). 'Picasso Speaks', *The Arts*, New York, May 1923, pp. 315–326.
74. Eugène Delacroix (1854/1948). *The Journal of Eugène Delacroix*, trans. Walter Pach (New York: Crown), p. 421.
75. Vincent Van Gogh, letter, quoted in John Russell, 'The Words of Van Gogh', *New York Review of Books*, 5 April 1979.
76. Samuel Palmer (1892). 'Shoreham Notebooks, 1824', quoted in Alfred Herbert Palmer, *The Life and Letters of Samuel Palmer, Painter and Etcher* (London: Sealey), p. 16.
77. Friedrich Nietszche (1999). *The Birth of Tragedy and Other Writings*, ed. Raymond Guess and Ronald Spears (Cambridge: Cambridge University Press), p. 113.
78. Daniel C. Dennett (2017). *From Bacteria to Bach and Back: The Evolution of Minds* (New York: WW Norton & Company), p. 345.
79. Michael S. A. Graziano, Arid Guterstam, Branden J. Bio, and Andrew I. Wilterson (2019). 'Toward a Standard Model of Consciousness: Reconciling the Attention Schema, Global Workspace, Higher-Order Thought, and Illusionist Theories', *Cognitive Neuropsychology*, 37, 155–172, 158.
80. Keith Frankish. 'The Consciousness Illusion', *Aeon*, 26 September 2019.
81. David J. Chalmers (2018). 'The Meta-Problem of Consciousness', *Journal of Consciousness Studies*, 25, 6–61, 26.
82. Nicholas Humphrey (1987). 'The Uses of Consciousness', James Arthur Memorial Lecture, American Museum of Natural History, New York. p. 19; reprinted in Nicholas Humphrey (2002). *The Mind Made Flesh* (Oxford: Oxford University Press), pp. 65–85, 82.
83. Stuart Sutherland (1984). 'Consciousness and Conscience', *Nature*, 307, 39, 233.
84. David Chalmers (2020). 'How Can We Solve the Meta-Problem of Consciousness? Reply', *Journal of Consciousness Studies*, 27, 201–226, 225.
85. David Chalmers (2018). 'How Can We Solve the Meta-Problem of Consciousness?', *Journal of Consciousness Studies*, 25, 6–61, 24.
86. Milan Kundera (1991). *Immortality*, trans. Peter Kussi (London: Faber & Faber), p. 225.

87. Anthony Kenny and Conrad Hal Waddington (1972). Extract from *The Nature of Mind*, The Gifford Lectures 1971/72 (Edinburgh: Edinburgh University Press).
88. William M. Marston, (1929). 'Consciousness', in *Encyclopaedia Britannica 14th Edition.*
89. David Balduzzi and Giulio Tononi (2009). 'Qualia: The Geometry of Integrated Information', *PLoS Computational Biology*, 5.8, e1000462, 1.
90. Samuel Taylor Coleridge (1834). *Biographia Literaria* (New York: Leavitt, Lord), p. 140.
91. Daniel Dennett (1991). *Consciousness Explained* (New York: Little Brown), p. 371.
92. Daniel Dennett (1996). *Kinds of Minds* (New York: Basic Books), p. 97.
93. William James (1990). *Principles of Psychology*, Vol 1 (New York: Henry Holt), p. 147.
94. George Wilhelm Friedrich Hegel (1812/2015). *The Science of Logic*, trans. George di Giovanni (Cambridge: Cambridge University Press), p. 322.
95. But see n 70 above, where I suggest an alternative scenario that could have resulted in sentience arising simultaneously in more than one modality.
96. I have cited evidence that Helen was cognitively conscious of her perceptual representations, even though she had no visual sensations: she *knew* what she saw (see n. 24). I would assume the same is true of other 'natural blindsighters' such as frogs. It's hard for us to imagine such a condition: 'what it's like' to have blindsight. But this may help. There's a perceptual phenomenon called 'amodal completion'. It's where you 'perceive' contours and surfaces for which there's no immediate evidence in the visual image, for example, the shape of an object that is partially occluded by another (Figure n. 96, left) or the outlines of the Kanisza triangle (Figure n. 96, right). If you were a natural blindsighter, perhaps everything you perceive would be at this level.
97. William James (1890). *Principles of Psychology*, Vol. 1 (New York: Henry Holt), p. 226.
98. ibid, pp. 321–323.

Figure n. 96.

99. The potential of raised temperature to facilitate positive feedback in the human brain is illustrated by what happens when by mischance temperature rises to fever level and the whole brain goes into epileptic seizure. In other animals, there is evidence of a beneficial effect of raised temperature on sensory physiology. Swordfish, though cold-blooded, can selectively raise the temperature of their eyes when they dive to great depths, with the result that their visual acuity increases by a factor of ten. K. A. Fritsches, R. W. Brill, and E. Warrant (2005). 'Warm Eyes Provide Superior Vision in Swordfishes', *Current Biology*, 15, 55–58.
100. There's possibly still more to this. If the attractors that are the vehicle for representing phenomenal properties are to work as we've suggested, it will have been important for their shape to be stabilized so that the representation would be consistent from one occasion to the next. But such stability might have been impossible to achieve in a brain whose fluctuating temperature meant that conduction velocities were varying all the time. Therefore warm-bloodedness could have been the essential precondition for the attractors to become reliable purveyors of phenomenal properties. If what it will be like for you to see red tomorrow will be what it was like for you to see red yesterday, you may have to thank your temperature-constant brain.
101. Michael Tye (2017). *Tense Bees and Shell-Shocked Crabs* (Oxford: Oxford University Press), p. 72.
102. Frans de Waal, quoted by Michael Gross (2013). 'Elements of Consciousness in Animals', *Current Biology*, 23, R981–R983, p. 983.

103. Michael Tye (2017), *Tense Bees and Shell-Shocked Crabs* (Oxford: Oxford University Press), p. 75.
104. There are accounts of octopuses engaging in object play, with a ball, for example. But none would appear to be examples of sensation seeking, and none involve social play.
105. Thomas Nagel (1979). *Mortal Questions* (Cambridge: Cambridge University Press), p. 2.
106. Paul Valéry (2021). *Notebooks*, in *The Idea of Perfection: The Poetry and Prose of Paul Valéry; A Bilingual Edition*, trans. Nathaniel Rudavsky-Brody (New York: Farrar, Straus and Giroux, 2021).
107. Michael Tippett (1979). 'Feelings of Inner Experience', in *In How Does It Feel?*, ed. Mick Csacky (London: Thames and Hudson), pp. 173–178, 175.
108. Quoted by Nicholas Spice (2019). 'Ne Me Touchez Pas', *London Review of Books*, 24 October, 20, 41.
109. Thomas Beecham (1961), quoted in *The New York Herald Tribune*, 9 March.
110. Charles T. Snowdon, David Teie, and Megan Savage (2015). 'Cats Prefer Species-Appropriate Music', *Applied Animal Behaviour Science*, 166, 106–111.
111. I focused on the difference between 'interest' and 'pleasure' in my early experiment on monkey aesthetics. Nicholas Humphrey (1972). 'Interest and Pleasure: Two Determinants of a Monkey's Visual Preferences', *Perception*, 1, 395–416.
112. Crickette Sanz (2009), quoted by Rebecca Morelle, '"Armed" Chimps Go Wild for Honey', *BBC News* Online, 18 March. http://news.bbc.co.uk/1/hi/sci/tech/7946614.stm
113. Bill Wallauer, 'Waterfall Displays', https://www.janegoodall.org.uk/chimpanzees/chimpanzee-central/15-chimpanzees/chimpanzee-central/24-waterfall-displays (accessed 10 May 2022).
114. Jane Goodall, commentary on 'Waterfall Displays', https://vimeo.com/18404370 (accessed 10 May 2022).
115. 'Genius Dog Takes Herself Sledding', 22 January 2018, https://www.youtube.com/watch?v=Pr2oknKroWQ (accessed 10 May 2022).

116. Claire Mitchell (2019). 'An Exploration of the Unassisted Gravity Dream', *European Journal of Qualitative Research in Psychotherapy*, 9, 60–71, p. 66.
117. Claudia Picard-Deland, Maude Pastor, Elizaveta Solomonova et al (2020) 'Flying Dreams Stimulated by an Immersive Virtual Reality Task', *Consciousness and Cognition*, Aug 83,102958.
118. A. F. McBride and D O. Hebb (1948). 'Behavior of the Captive Bottle-Nose Dolphin, *Tursiops truncatus*', *Journal of Comparative and Physiological Psychology*, 411, 111–123.
119. Erica Jong (1973). *Fear of Flying* (New York: Holt, Rinehart and Winston).
120. Christina Hunger (2021). *How Stella Learned to Talk: The Groundbreaking Story of the World's First Talking Dog* (New York: William Morrow).
121. Alex Kirby (2004). 'Parrot's Oratory Stuns Scientists', *BBC News Online*, updated 1 May 2007, http://news.bbc.co.uk/2/hi/science/nature/3430481.stm (accessed 10 May 2022).
122. Milan Kundera (1991). *Immortality*, trans. Peter Kussi (London: Faber and Faber, 1991), p. 50.
123. Dennett is less willing than I am to credit dogs with having a self-conscious phenomenal self.

    > Dogs presumably do not think there is something it is like to be them, even if there is. It is not that a dog thinks there isn't anything it is like to be a dog; the dog is not a theorist at all, and hence does not suffer from the theorists' illusion. The hard problem and meta-problem are only problems for us humans, and mainly just for those of us humans who are particularly reflective. In other words, dogs aren't bothered or botherable by problem intuitions. Dogs—and, for that matter, clams and ticks and bacteria—do enjoy (or at any rate benefit from) a sort of user illusion: they are equipped to discriminate and track only some of the properties in their environment.

    Daniel Dennett (2019). 'Welcome to Strong Illusionism', *Journal of Consciousness Studies*, 26, 48–58, 54.
124. Claudia Fugazza, Péter Pongrácz, Ákos Pogány, Rita Lenkei, and Ádám Miklósi (2020). 'Mental Representation and Episodic-Like Memory of Own Actions in Dogs', *Science Reports*, 10, 10449.

125. William Hazlitt (1805). *An Essay on the Principles of Human Action*, (London: J Johnson) p. 3.
126. Shoji Itakura (1992). 'A Chimpanzee with the Ability to Learn the Use of Personal Pronouns', *Psychological Record*, 42, 157–172.
127. Nicholas Humphrey (1978). 'Nature's Psychologists', 1977 Lister Lecture of the British Association for the Advancement of Science [short version], *New Scientist*, 29 June 1978, 900–904, 901.
128. David Premack and Guy Woodruff (1978). 'Chimpanzee Problem-Solving: A Test for Comprehension", *Science*, 202, 532–535.
129. David Premack and Guy Woodruff (1978). 'Does the Chimpanzee Have a Theory of Mind?", *Behavioral and Brain Sciences*, 4, 515–526. Interestingly, the authors mistakenly list their previous *Science* paper as 1975 rather than 1978. Guy Woodruff (personal communication, 2020) has told me that when they embarked on this set of experiments with Sarah they had no intention of studying 'theory of mind'. It only occurred to Premack to interpret their findings in these terms a year later (after reading my lecture?).
130. Celia M. Heyes (1998). 'Theory of Mind in Nonhuman Primates', *Behavioral and Brain Sciences*, 21, 101–148.
131. Daniel Dennett (1987). *The Intentional Stance* (Boston, MA: MIT Press), p. 17.
132. Nicholas Humphrey (1980). 'Nature's Psychologists' [full version] in *Consciousness and the Physical World*, ed. B. Josephson and V. Ramachandran (Oxford: Pergamon), pp. 57–75, 73–74.
133. Florence Gaunet (2008). 'How Do Guide Dogs of Blind Owners and Pet Dogs of Sighted Owners (*Canis familiaris*) Ask Their Owners for Food?', *Animal Cognition*, 11, 475–483, 482.
134. Emmanuel Levinas (1990). 'The Name of a Dog, or Natural Rights', in *Difficult Freedom: Essays in Judaism*, trans. Sean Hand (Baltimore, MD: Johns Hopkins University Press), pp. 151–153.
135. ABC News, "Gorilla Carries 3-Year-Old Boy to Safety in 1996 Incident', https://www.youtube.com/watch?v=puFCuMacoVk (accessed 10 May 2022).
136. Helen Macdonald (2021). *Vesper Flights: New and Collected Essays* (London: Vintage), p. 155.
137. Mylene Quervel-Chaumette, Viola Faerber, Tamas Farago, Sarah Marshall-Pescinil, and Friederike Range (2016). 'Investigating

Empathy-Like Responding to Conspecifics' Distress in Pet Dogs', *PLoS ONE* 11,4, e0152920.

138. Nobuya Sato, Ling Tan, Kazushi Tate, and Maya Okada (2015). 'Rats Demonstrate Helping Behaviour Toward a Soaked Conspecific', *Animal Cognition*, 18, 1039–1047.

139. Stanley Wechkin, Jules H. Masserman, and William Terris (1964). 'Shock to a Conspecific as an Aversive Stimulus', *Psychonomic Science*, 1, 47–48, 237.

140. There was a recent report in the Cambridge local paper about a muntjac that was hit and badly injured by a car. It went on screaming for hours—until it was dispatched by a vet. Presumably it was screaming *in pain*. But *why*? In what possible circumstances could such behaviour be biologically adaptive? I don't have answers.

141. Charles Darwin (1871). *The Descent of Man, and Selection in Relation to Sex* (London: John Murray), p. 147.

142. A. V. Hill, letter to *The Spectator*, 18 May 1945.

143. Monkey video, https://cdn.theguardian.tv/mainwebsite/2014/12/22/141222Monkey_desk.mp4 (accessed 10 May 2022).

144. An additional factor, for humans, could have been the need to counter thoughts about suicide. I discuss why humans, alone among animals, may develop a death wish, and the role of the phenomenal self as a psychological defence against it, in Nicholas Humphrey (2017). 'The Lure of Death: Suicide and Human Evolution', *Philosophical Transactions of the Royal Society, B*, 373, 20170269.

145. Peter Walling and Kenneth Hicks have examined the electrical activity of the central nervous system of a range of different animals, looking for evidence of attractors in the EEG. Although this may have only a distant bearing on the sensorimotor attractors that underlie phenomenal consciousness, their findings are interesting and germane. The figure below shows '"the highest recorded attractor dimension plotted against the approximate time of each species" first appearance (billions of years ago)'. Note that mammals have much higher dimensional attractors than fish or octopuses (no birds included). See Peter T. Walling (2020). 'An Update on Dimensions of Consciousness', *Baylor University Medical Center Proceedings*, 33,1, 126–130.

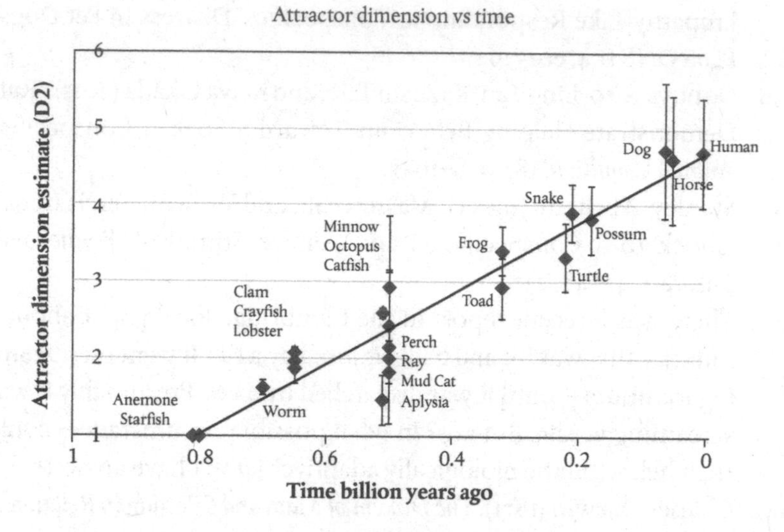

Figure n. 145 Highest recorded attractor dimension plotted against the approximate time of each species (billions of years ago)

From Peter T. Walling (2020). 'An Update on Dimensions of Consciousness', *Baylor University Medical Center Proceedings*, 33,1, 126–130.

146. Daniel Dennett, letter to Toby Mundy, 2021.
147. Some popular accounts are certainly overly romantic. The 2020 Netflix film *My Octopus Teacher* purports to be a record of how an octopus befriended a man. Like the film, *ET*, it's a beautifully crafted story. How remarkable, indeed, that an octopus—a species that never makes friends with any of its own kind—should have formed this bond with a human. However, as the producer's blog makes clear, it's largely a fiction, a work of art assembled in the cutting room.
148. Peter Godfrey-Smith (2020). *Metazoa: Animal Minds and the Birth of Consciousness* (London: Collins), p. 142.
149. The one example of octopuses' behaviour that gives me pause—and might suggest their having a travelling sense of self—is evidence that they carry coconut shells to a new location where they can then hide in them. 'We repeatedly observed soft-sediment-dwelling octopuses carrying around coconut shell halves, assembling them as a shelter only when needed.' See Julian K. Finn, Tom

Tregenza, and Mark D. Norman (2009). 'Defensive Tool Use in a Coconut-Carrying Octopus', *Current Biology*, 19, R1069–R1070.

150. Godfrey-Smith (2020). *Metazoa: Animal Minds and the Birth of Consciousness* (London: Collins), p. 146.
151. ibid, p. 109.
152. Susan Schneider and Edwin Turner (2017). 'Is Anyone Home? A Way to Find Out If AI Has Become Self-Aware', *Scientific American Blog Network*, 19 July.
153. Thomas Mann, quoted in Edward R. Murrow (ed.) (1952). *This I Believe: The Living Philosophies of One Hundred Thoughtful Men and Women in All Walks of Life* (New York: Simon & Schuster).
154. Mary Oliver (2020). 'Do Stones Feel?', in *Devotions: The Selected Poems of Mary Oliver* (New York: Penguin).
155. A case in point is the Animal Welfare (Sentience) Bill in the United Kingdom. This was amended in 2021 to cover octopuses and lobsters (among others), following a report by a working party at The London School of Economics. Jonathan Birch, Charlotte Burn, Alexandra Schnell, Heather Browning, and Andrew Crump (2021). 'Review of the Evidence of Sentience in Cephalopod Molluscs and Decapod Crustaceans' (London: LSE Consulting). The authors do not attempt to analyse the concept of sentience but simply state at the outset that 'to be sentient is simply to *have* feelings'. From there on, they use the term 'feeling' to mean any kind of valenced mental state, not restricted to sensations that have phenomenal properties. Thus they leave it to readers to draw the analogy—as they naturally will do—with *human feelings*. They go on to provide a thorough review of the evidence that, for instance, lobsters respond adaptively to 'painful stimuli', but what they fail at any point to establish is that lobsters are conscious of pain in the way that humans are, let alone that lobsters possess a phenomenal self that they mind about.
156. Nicholas Humphrey (1979). 'New Ideas, Old Ideas', *London Review of Books*, Vol. 1, No. 4, 6 December.

INDEX

Adorno (philosopher)
  music without sound 16
Agresara (spirit guide) 22
'AI Consciousness Test' 209
Animal Welfare (Sentience)
    Bill (UK) n.155
Aristotle
  animal compassion 192, 199
art, as 'improvement' on
    nature 116–8
attractor states
  created by feedback 109–11, 140
  delay differential equations
    and n.69
  dimensions of, in EEG n.145
  'ipsundrum' as special case
    of 111–12

Balinspittle, moving statue 27–9
Bateson, Gregory
    (anthropologist) 216
Beecham, Thomas (musical
    conductor) 163
Bentham, Jeremy (philosopher)
  animal suffering 9, n.5
Bertrand, Mireille (primatologist)
  taming of Helen 42
blindsight
  absence of 'self' in 48, 52
  after early blindness (?) 50–3
  as cognitively conscious
    vision n.24, n.96
  as perception without
    sensation 46
  compared to amodal
    completion n.96
  failure to understand vision in
    others 75, 188
  frog vision as 46, 48
  in human patients 45–6
  in monkey 40–5
Block, Ned (philosopher)
  'phenomenal' and 'access'
    consciousness 7
Borman, Frank (astronaut) x, 212
brain, user-illusion 119–120, 123
  qualia-coding of sensory
    modalities 120–2
Brindley, Giles (physiologist)
  research on phosphenes 16
Broad, Charles (philosopher)
  psychical research 21–3
  views on survival of mind after
    death 25–6
Broks, Paul (neuropsychologist)
  loss of self in Cotard's
    syndrome n.30
Burns, Robert (poet) 96

Cartesian diver 140
Chalmers, David (philosopher)
  miscellaneous 3, 7, 120,
    126, 128
chimpanzees
  honey gathering 167
  masturbation 170
  mental time-travel 181
  mind-reading 185–7
  mirror-test 178
  self-identity 182
  waterfall display 167–8
Clayton, Nicola (comparative
    psychologist)
  mental time-travel by birds 181
Clark, Tom (philosopher)
  mental representation n.53

Coleridge, Samuel Taylor (poet)
 'touch me mother' 96 , 115
 rule for understanding others' beliefs 138
colours
 monkeys' affective responses to 54–66
compassion
 evidence of, in animals 192–6
consciousness, cognitive and phenomenal
 functional overview 3–9

Darwin, Charles
 eye, evolution of 102
 compassion, human capacity for 196
Delacroix, Eugene (painter) 117
Dennett, Daniel (philosopher)
 author's relation to 27
 behavioural dispositions as key to consciousness 65–6, 89, 93–4
 brain user-illusion 119
 dogs' lack of philosophical reflection n.123
 grades of consciousness 140–1
 'hard question' 89
 intentional stance 187
 phenomenal quality as illusion 80, 139
Descartes, René
 animals as machines xi
 idea of God, inadequate cause for 84–5
dogs
 death awareness 198
 empathy 193, 195
 toboggan play 169
 mind-reading 190–1
 music, response to 164
 self-identity 177–9
 time-travel 180
dreams of flying 170
Dunbar, Robin (anthropologist)
 Dunbar's number 74, n.37

efference copy 108, n.68

Fodor, Jerry (philosopher)
 hard problem xiv
Fossey, Dian (ethologist)
 author's visit to 67–70
 personal character of 69–70
Frankish, Keith (philosopher)
 illusionism 80, 119

Godfrey-Smith, Peter (philosopher)
 mind of octopus 204–6
Goff, Philip (philosopher)
 panpsychism 86, 88, 95, 137
Golden Rule of ethics 215
Goodall, Jane (ethologist) 67, 168
gorillas, mountain
 as 'natural psychologists' 72–4
 author's observations of 68–73
 skulls and brain size 68–9
Graziano, Michael (psychologist)
 brain user-illusion 120
Groves, Colin (anthropologist)
 gorilla skulls 68–9

Hegel, (philosopher)
 discontinuities in evolution 142
Helen (monkey)
 recovery of vision after removal of visual cortex 40–5
 video n.19
Heyes, Celia (comparative psychologist)
 critique of mind-reading 186
Hume, David (philosopher)
 projection of feelings 82
 sensations as foundation for self 114–116

Huxley, Thomas Henry
'soul of frog' 91
hydrancephalic children 92–3

Integrated Information Theory (IIT) 137–138
inverted spectrum 97–9
'inner eye' 123
ipsundrum (attractor state)
as vehicle for phenomenal properties 110–112, 147–148

Jackson, Frank (philosopher)
'Mary' thought-experiment 94
James, William (psychologist)
grades of consciousness 142
individual selfhood 151
Jong, Erica (author) 170

Kenny, Anthony (philosopher) 136
Klee, Paul (painter) 116
Knock, miraculous apparition at 29–30, 90
'Knowledge argument' 94–100
Koch, Christof (neuroscientist)
neural correlate of consciousness (NCC) 87
integrated information theory 138
Kundera, Milan (author)
phenomenal selfhood 133, 156
face as 'serial number of species' 178

life after death, reasons for belief in 200–1
Lloyd, Dan (philosopher)
'transparent' theory of consciousness 84, 86
Locke, John (philosopher)
inverted spectrum 99

MacDonald, Helen (poet)
swan's sympathy for 193
Madingley (Sub-Department of Animal Behaviour) 41–42, 60, 66
Mann, Sargy (blind painter)
hallucination of colour sensations 83–84
Mann, Thomas (novelist)
'man' as goal of creation 213
masturbation
phenomenal self and 133, 170–172
'Mary' thought-experiment 94–100
McGinn, Colin (philosopher)
explanatory gap 85, 88
Miller, Geoffrey
consciousness and courtship 124
mind-reading 5, 74–6, 119, 123–4, 126, 128, 186, 191, 194
see also theory of mind
Montero, Barbara (philosopher)
pain amnesia n.63
Morgana, Aimeé
studies with talking parrot (N'kisi) 177
musical appreciation
as sensation for sensation's sake 162–3
by animals 164–7

Nagel, Thomas (philosopher)
phenomenal experience as intrinsically positive 160–1
'Nazis, Animal Protection Law 196–197
neural correlate of consciousness (NCC) 87, 136, 138
neural correlate of representing consciousness (NCRC) 88, 138
Newton, Isaac
'Newton's rule' 154, 156
research on phosphenes 6–8
Nietzsche, Friedrich (philosopher)
art as metaphysical extension of nature 118
N'Kisi (talking parrot) 177

octopuses
  sentience of 204–205, n.147, n.149
Oliver, Mary (poet)
  'Do Stones Feel?' xiii, 214
orgasm
  as body music 172
  as pure sensation without perception 131–3
  phenomenology of 131–3, 170–3

pain, phenomenal
  no simple relation to pain behaviour 154–6
Palmer, Samuel (painter) 117
panpsychism
  vacuousness of xii, 86, 93, 95, 139
phenomenal surrealism n.41
Picasso, Pablo (painter) 116
play
  role in development of self 159
  seeking sensation for sensation's sake 160
position sense (proprioception)
  as pure perception without sensation 131
Premack, David and Guy Woodruff
  chimpanzee's 'theory of mind' 185, 187–188, n.129

qualia-coding
  sensory modalities and 120–2

Reid, Thomas (philosopher)
  'dual province of senses' 19–20, 46–7, 56, 82
  anticipation of blindsight 47
representation (mental)
  features of 78–79
  Tom Clark on n.53
robotic sentience
  possibility of 208
  tests for 209–10
Russell, Bertrand (philosopher)
  postulating as theft 86

Schneider, Susan and Edwin Turner
  tests for consciousness in a machine 209–10
self
  as subject of conscious mental states 4
  unity of 4–5
  phenomenalization of 108–12
self (phenomenal)
  as work of art 116–8
  attribution to others 119
  mind-reading and 118–24
  mysterious properties of 118
  sensations as basis for 114–6
  sustenance of 158–174
  uses of 175–201
sensations
  evolutionary history of 105–113
  phenomenal character of 78–81
sensation and perception
  sensation as representation of 'what's happening to me' 79
  perception as representation of 'what's happening out there'
  see also Reid, 'dual province'
sentience
  definition 1–2
  diagnostic criteria for 146, 153, 157
sexual selection 104, 124
Simpson, Joe (climber)
  pain as affirmation of self 93, 133
'social function of intellect' 74
Solms, Mark (neuropsychologist) 92
soul
  as cultural elaboration of phenomenal self 118, 201
superior colliculus (optic tectum)
  author's recording in monkeys 34–8, 46
Sutherland, Stuart (psychologist) 125–6

theory of mind 183–91
  see also mind-reading, brain user-illusion

Tippett, Michael (composer)
music as nourishment for soul 162
Tononi, Giulio (neuroscientist)
integrated information theory 137–8
Tye, Michael (philosopher)
inference from human to animal consciousness 154, 156

Van Gogh, Vincent (painter) 117
Vallortigara, Giorgio (neuroscientist)
efference copy and consciousness n.68

Waddington, C.H. (biologist) 136
Wallace, Alfred Russel (naturalist)
brain as inadequate cause for consciousness 86
Walling, Peter and Kenneth Hicks
EEG attractors across animal kingdom n.145
warm-bloodedness
autonomous selfhood and 150–1
evolution of 149–50
nerve conduction speed, effect on 152, n.99
Weiskrantz, Lawrence (neuropsychologist)
author's relation to 32, 41
blindsight, discovery of 45
visual cortex, research on 32–3
Whitaker, Hugh (spirit communicator)
author's visit to in Elba 22–25